REFRACTORY SEMICONDUCTOR MATERIALS

TUGOPLAVKIE ALMAZOPODOBNYE POLUPROVODNIKI

ТУГОПЛАВКИЕ АЛМАЗОПОДОБНЫЕ ПОЛУПРОВОДНИКИ

REFRACTORY SEMICONDUCTOR MATERIALS

Yu. V. Shmartsev • Yu. A. Valov • A. S. Borshchevskii

Editors
N. A. Goryunova • D. N. Nasledov
A. F. Ioffe Physicotechnical Institute
Academy of Sciences of the USSR

Translated from Russian by
Albin Tybulewicz
Editor, Physics Abstracts, London

Springer Science+Business Media, LLC 1966

ISBN 978-1-4899-4769-7   ISBN 978-1-4899-4767-3 (eBook)
DOI 10.1007/978-1-4899-4767-3

The Russian text, corrected and updated by the authors for the English
edition, was originally published by Metallurgiya Press in Moscow in 1964.

Library of Congress Catalog Card Number 65-27346

© 1966 Springer Science+Business Media New York
Originally published by Consultants Bureau in 1966.
Softcover reprint of the hardcover 1st edition 1966

A Division of Plenum Publishing Corporation
227 West 17 Street, New York, N. Y. 10011

PREFACE TO AMERICAN EDITION

We learned with pleasure that our work is to be published in the USA, a country in which fundamental investigations of the physics of semiconductors and semiconducting devices abound and where semiconducting devices have found a wide range of practical application. This gives us hope that our review of the "difficult" but very promising semiconducting materials may, at least to some extent, stimulate work on the technology of these materials and on their physical properties. Undoubtedly, these efforts will finally lead to the production of new semiconducting devices, some of which will be based on principles that exploit the characteristic features of the band structure of these new semiconducting materials.

We are certain that our monograph is not free from errors and deficiencies. Therefore, we would welcome comments from any of our readers.

Yu. V. Shmartsev, Yu. A. Valov, and A. S. Borshchevskii

A. F. Ioffe Physicotechnical Institute,
Academy of Sciences of the USSR
February 1966

PREFACE

Semiconductors belong to a new branch of technology that has developed mainly in the last 10-12 years. They are vitally important in the realization of current plans for the extensive development of the national economy because their use is rapidly transforming and extending electronic engineering, automation, and power engineering.

Consequently, it is necessary to prepare new semiconducting materials possessing a range of specified electrophysical properties.

In some cases, the established semiconducting materials, such as germanium, silicon, or indium antimonide, cannot satisfy the ever growing range of requirements and therefore a large amount of research effort is being directed toward the preparation and study of new semiconducting compounds. Already, new semiconducting devices, such as masers, tunnel diodes, and radiation detectors, are being made of semiconducting materials which have found no practical application before.

The book deals with the synthesis and properties of a large group of refractory semiconducting materials, mainly of the $A^{III}B^V$ type.

This is the first attempt to present systematically the data on these materials. The authors have incorporated in the book their own work, carried out in the last decade, on the synthesis and properties of some of these materials.

The first chapter of the book deals with the general problems of the physics and chemistry of semiconductors. The second chapter describes the properties of a number of materials.

The information gathered together in the book suggests that the semiconducting substances concerned are very promising as materials for the fabrication of known types of device (diodes, photoresistors, transistors) and for the fabrication of devices with basically new properties.

The characteristic common to all the materials described in the book is the relatively large forbidden band width. Consequently, these compounds exhibit intrinsic conduction only at elevated temperatures which favors their use in devices working up to 1000°C.

We hope that the present book will promote the quicker and wider adoption of the materials described in it.

N. A. Goryunova and D. N. Nasledov

PUBLISHER'S NOTE

The following Soviet journals cited in this book are available in cover-to-cover translation:

Russian Title	English Title	Publisher
Doklady Akademii Nauk SSSR	Soviet Physics — Doklady	American Institute of Physics
Fizika tverdogo tela	Soviet Physics — Solid State	American Institute of Physics
Izvestiya Akademii Nauk SSSR: Seriya fizicheskaya	Bulletin of the Academy of Sciences of the USSR: Physical Series	Columbia Technical Translations
Zhurnal eksperimental'noi i teoreticheskoi fiziki	Soviet Physics—JETP	American Institute of Physics
Zhurnal neorganicheskoi khimii	Russian Journal of Inorganic Chemistry	The Chemical Society (London)
Zhurnal obshchei khimii	Journal of General Chemistry of the USSR	Consultants Bureau
Zhurnal tekhnicheskoi fiziki	Soviet Physics — Technical Physics	American Institute of Physics

CONTENTS

INTRODUCTION

Semiconductors occupy a special place in the history of science and technology. Semiconducting devices were first announced only a couple of decades back, but now the variety of semiconducting materials available and the number and range of devices fabricated from them are enormous.

Up to 1950, the term "semiconductor" was associated primarily with such substances as cuprous oxide, selenium, and germanium. The properties of each of these materials restricted their application in many branches of technology because they did not satisfy requirements for certain combinations of properties. The early fifties were marked by the urgent search for new semiconducting materials.

The outstanding Soviet scientist, Academician A. F. Ioffe, suggested that semiconducting properties were connected with chemical structure. On his initiative, the Physicotechnical Institute of the USSR Academy of Sciences in Leningrad began a detailed study of gray tin, which is the chemical analog of the well-known semiconductor, germanium. This investigation, carried out by N. A. Goryunova [1], showed that gray tin is, like germanium, a typical semiconductor. During her studies of the charging of white tin and her attempts to convert it into the β-modification, Goryunova discovered that indium antimonide InSb and cadmium telluride CdTe, deposited on the surface of white tin, gave rise to, and greatly accelerated the process of, the transformation of white into gray tin. This indicated that the chemical binding in these substances was similar. It was suggested, and soon confirmed, that the "chargeable" substances, such as gray tin and germanium, are semiconductors.

In 1952, a German scientist, Welker, reported [2] that some binary chemical compounds, prepared from elements of groups III and V (the $A^{III}B^V$ compounds) are typical semiconductors.

The decade following 1952 saw the rapid development of the semiconductor technology. Hundreds of papers appeared on new semiconducting materials, the number of which grew rapidly. This was principally due to studies of the relationships governing the formation of substances which were analogs of silicon and germanium. The relationships governing the changes of semiconducting and other properties were also investigated. As a result of this work, it is now possible to approach scientifically the selection of new materials and to predict the principal and most characteristic properties of new semiconducting materials.

It is evident from the numerous theoretical and experimental studies that, among the multitude of semiconducting materials, one's attention has become focused on the elements of the subgroup IV^b (germanium, silicon, diamond), the only known compound of the $A^{IV}B^{IV}$ type — silicon carbide (SiC), and compounds of the $A^{III}B^V$ type. This situation is not accidental. The most powerful stimulus in the investigation of semiconductors has been the invention of semiconducting devices — diodes and transitors — capable of replacing electron tubes in radio circuits. At present, radio engineering is the main user of the numerous types of semiconducting device.

In 1958, semiconducting diodes and transitors were being manufactured by 76 firms in various countries. The number of devices produced was about 1 billion. The high reliability, mechanical strength, exceptionally small size, and low power consumption of semiconducting devices have led to their ever increasing use in electronic circuits.

At present, the most promising electronic engineering materials are those which have been little used so far: silicon carbide, diamond, and some "refractory" compounds of the $A^{III}B^V$ type. The latter have been dealt with in books and review articles [3] and some of the $A^{III}B^V$ compounds (InSb, GaAs) are already widely used in industry; this is because they are relatively easy to make. However, among the numerous compounds of the

1

$A^{III}B^{V}$ type there is a group of semiconductors which have been studied very little, although they should be very interesting from the point of view of their potential applications in technology. They are the refractory materials: nitrides, phosphides, and arsenides of boron and aluminum, together with indium and gallium phosphides. Similar properties are exhibited by diamond and silicon carbide. All these substances (which need further study) form the subject of the present book.

The sparse information on the methods of preparation, properties, and applications of these semiconductors is scattered among various scientific journals and in patents. Having gathered together this information and compared the published data with their own, the authors of the present book came to the conclusion that the utilization of these materials in power and radio engineering may give rise to devices capable of operation in an extremely wide range of conditions: high and low temperatures, corrosive environments, i.e., under conditions in which there is a growing interest.

In order to be able to mass-produce semiconducting devices, it is necessary to have the required semiconducting materials in sufficient amounts. Therefore, the present book is intended for metallurgists, who should be the first to produce new semiconducting materials on which the manufacture of semiconducting devices can be based.

In the first part, information is given on the main technical methods used in the synthesis and growth of single crystals of the semiconducting materials discussed in the second section, which summarizes all the known data on the preparation, properties, and possible applications of diamond, silicon carbide, nitrides, phosphides, and arsenides of boron and aluminum, as well as of indium and gallium phosphides.

The authors would be completely satisfied if their work contributed, in some degree, to the development of the metallurgy of new semiconducting materials and thus to the progress of all semiconductor technology.

GENERAL TECHNOLOGY FOR PREPARATION
OF SEMICONDUCTING MATERIALS

General Information

Since the electrical properties of a semiconductor are extremely sensitive to defects in its crystal structure and to the presence of impurities, the preparation of semiconducting crystals from raw materials should be carried out in such a way as to control these imperfections. The control of structural defects is important only in the last stage of the process of preparing a semiconductor, i.e., the growth of single crystals. But the control of impurities is important in the many earlier stages of the preparation.

The impurities that we speak of here are not the alloying admixtures which are deliberately introduced into a semiconducting material to give it the required electrical properties, but the foreign substances which we shall call "initial" and "process" impurities. The initial impurities are those present in the original raw materials. The process impurities are those which enter a semiconducting material during its preparation.

Since the required electrical properties can be obtained by alloying a semiconducting material in the final stage, it is desirable to reduce the content of the initial and process impurities to a minimum — in other words, to achieve the highest possible purity. *

The purification of any material presents a problem which grows in complexity as the purity requirements are made more stringent. Since the impurity content of a semiconductor has to be reduced to an extremely low level (of the order of 10^{-6} % or less), it will be useful to consider the general basis of the methods used to prepare semiconducting materials.

The process of preparing a semiconducting material may be represented by the scheme shown in Fig. 1 where the boxes represent certain states of the material and the arrows represent technological processes.

If the constituents of the substance to be synthesized are difficult to purify, it is preferable to use as the initial materials suitable compounds which contain these constituents but are easy to purify. Thus, in the preparation of silicon it is not desirable to use technical-grade silicon as the initial raw material (I). It is much more convenient to use those silicon compounds ($SiCl_4$, SiI_4, SiH_4, and others) which have low melting or boiling points and which can be purified by the classical physicochemical methods (distillation, sublimation, recrystallization, etc.), much more effectively than technical-grade silicon. The same applies to semiconducting compounds. Using this approach, we can obtain spectroscopically pure materials (II) from which the constituents (III) of the required compound can be obtained by certain chemical reactions.

The preparation of spectroscopically pure components (III) does not eliminate the need for further purification of the synthesized compound. Since many semiconducting compounds have relatively high melting points, it is desirable to purify them at this stage by crystallization methods (zone melting, repeated pulling of a crystal from the melt). This purification may either precede the growing of single crystals or may form part of the process of growing.

The selection of a suitable preparation process is of basic importance in the attempt to increase the purity of a semiconducting material. Thus, in the preparation of certain complex semiconductors (for example,

---

*It is necessary to point out that in industry, where the economic factors are important, the problem of purification is approached in a different way: an attempt is made to obtain that degree of purity of a material which would give the optimum relationship between the quality and cost of semiconducting devices made from the material.

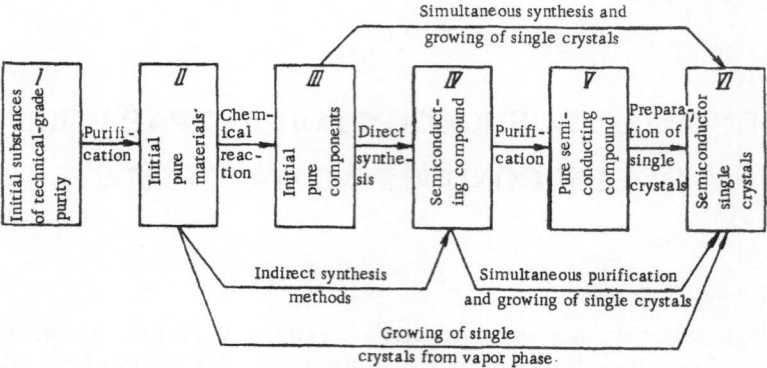

Fig. 1. Schematic representation of the production of semiconductor single crystals from raw materials.

GaAs) there is a recent tendency to carry out simultaneously the purification of the initial components, the synthesis, and the growing of single crystals. The whole process is carried out in a single stage, which is naturally much more complex than the usual process of the synthesis or growing of single crystals. However, such a method allows us to increase the degree of purity of the material, for example, by the removal of oxygen from gallium, indium, arsenic, etc.

In considering new initial materials and rational technological processes from the point of view of increasing the purity of semiconducting materials, it is always necessary to take account of crucible materials as potential sources of contamination. Because of this, the fusion methods which do not require the use of a crucible have assumed great importance.

The foregoing remarks are quite general. However, the technology of the preparation of refractory semiconducting materials is at present developed so little that a reminder of these general aspects may be useful in the development of new technological processes.

Synthesis of Semiconducting Compounds

The discovery of semiconducting properties in various compounds was followed by the establishment of a new technology: that of synthesizing of complex semiconductors, which did not exist when only the elemental semiconductors were known. By synthesis, we mean the chemical reaction which gives rise to the required compound from the initial materials. The synthesis of some semiconductors is relatively simple but that of others is fraught with difficulties.

It is necessary to distinguish clearly between the synthesis of a compound and its preparation in the form of an ingot. These two processes can be simultaneous for many compounds. All that is necessary is for the reaction to take place at a temperature higher than the melting point of the compound to be formed. Then the latter will be produced in the liquid state and will form a solid ingot on cooling. However, the synthesis of a compound is usually also possible at temperatures below its melting point. Then, the physical state of the synthesized compound is governed by the properties of the actual components taking part in the reaction. In those cases when the compound is formed from refractory or volatile components (BP, BAs, etc.) the synthesis involves a heterogeneous reaction of a vapor and a solid. Then the rate of the diffusion processes is low and the physical state of the reaction product is governed to a considerable extent by the physical state of

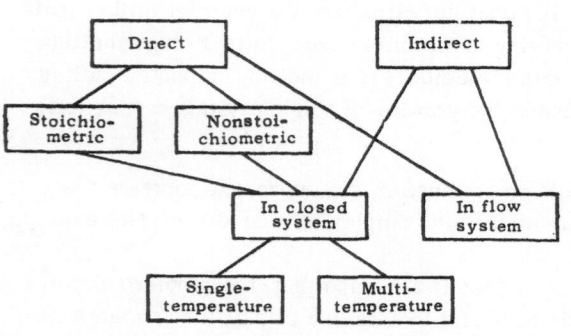

Fig. 2. Classification of the main methods of synthesizing semiconducting compounds.

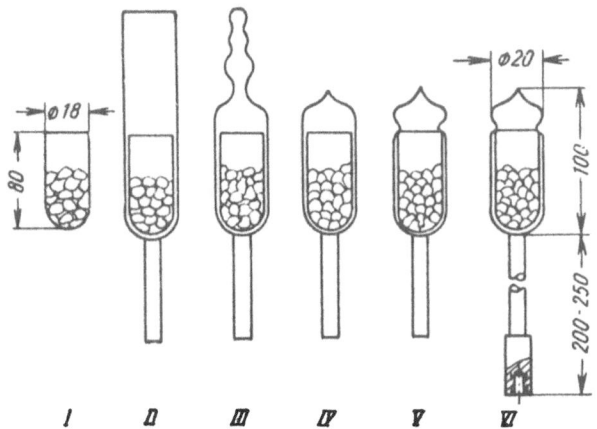

I   II   III   IV   V   VI

Fig. 3. Sequence of operations in the preparation of an
ampoule for synthesis: I) placing the charge in a quartz
test tube; II) placing the test tube in the ampoule; III)
making a constriction; IV) sealing the ampoule; V) making a constriction of the ampoule in order to fix the test
tube; VI) attachment of a quartz tube, which in turn is
attached to a metal rod (V and VI are necessary in the
synthesis with vibrational mixing).

the refractory component (thus, if we use boron powder to synthesize BP, the product of synthesis is usually a
powder). When one of the components of a compound (A^{III}) has a low melting point and the other (B^V) is volatile, then, below the melting point of the compound, the reaction takes place between a liquid and a vapor
(AlAs, InP, etc.). In this case, the synthesis involves the crystallization of the required compound from a liquid
solution of the volatile component in the low-melting-point component (for example, phosphorus in indium).
The product of a synthesis of this type may be a porous crystalline aggregate, consisting of crystallites precipitated from the solution as it becomes supersaturated. However, in those cases when the initial charge does
not consist of components taken in the stoichiometric ratio but has an excess of the metal (A^{III}) we can also
obtain the required compound in the form of small single crystals embedded in the excess of the metal component.

Classification of the Methods of Synthesis

The methods of synthesis may be classified according to the type of chemical reaction or according to
the technological features of this reaction. Figure 2 shows the classification scheme of the main methods.

Depending on the nature of the initial substances, all the methods can be divided into direct or indirect
synthesis. The direct methods are those in which the initial substances are the constituents of the required
compound (for example, the preparation of GaP from gallium and phosphorus). Therefore, the characteristic
feature (and the advantage) of the direct methods of chemical synthesis is that the reaction product consists
only of the required compound, if the initial substances are taken in the stoichiometric ratio (provided the reaction is carried out in a closed chamber and the reaction goes only in the forward direction). The direct
methods can also be nonstoichiometric when the initial charge contains some excess of the metallic component.

The indirect methods are those in which at least one of the initial substances is not a constituent of the
required compound (for example, the preparation of GaP from Zn_3P_2 and gallium). In this case the reaction
products include not only the required compound but some secondary product (zinc, in our case).

The methods of synthesis can be divided into several main groups in accordance with the technological
process involved. The synthesis may be carried out in a closed chamber, for example, in a sealed ampoule,
or we can use continuous flow systems, in which one of the reacting substances (in the gaseous state) flows
through a reaction chamber containing the second reagent.

The synthesis in a flow system requires much more complicated equipment; moreover, we need a great excess of the gaseous substance compared with its stoichiometric amount. Nevertheless, in some cases, the direct synthesis is carried out in a flow system (for example, the direct synthesis of nitrides). The flow systems are successfully employed in the preparation of crystals of compounds from their constituents.

The synthesis methods carried out in closed chambers can be divided into processes taking place at one temperature and those taking place at two (or more) temperatures. In the single-temperature process all parts of the ampoule are kept at the same temperature. In a multi-temperature process, different parts of the ampoule are maintained at different temperatures. The more usual are the two-temperature processes, in which one part of the working chamber is kept at a higher temperature than the other part.

To carry out a synthesis in a closed system, one uses ampoules which are generally made of quartz. A charge is placed in a prepared (washed, heated, etc.) ampoule, which is then provided with a constriction. An ampoule prepared in this way is then evacuated (or filled with an inert gas at a desired pressure) and sealed. Figure 3 shows the sequence of operations in the preparation of an ampoule for undertaking a synthesis.

If the charge includes a component which reacts with quartz at high temperatures (aluminum, beryllium, etc.), the charge is placed in a special test tube, which is then inserted in the ampoule, the test tube being made of a material which does not react with the initial components.

If the quartz fractures when the sample solidifies, a quartz test tube (or a boat, if the ampoule is horizontal) is used to contain the charge in the quartz ampoule. In the absence of such a test tube, the ampoule may fracture during the solidification and the sample oxidize. If a test tube is used, only it and not the ampoule fractures and the sample remains sealed.

We shall now consider the methods of synthesis which are widely used in the preparation of semiconducting compounds.

Direct Synthesis in a Closed Chamber at One Temperature

The charge is placed in a quartz test tube and the constituents of the compound are taken in their stoichiometric proportions. The test tube is evacuated, sealed, and placed in a furnace where it is heated uniformly. The temperature is increased in accordance with a schedule which depends on the characteristic features of the materials present in the charge. After reaching the maximum temperature, the ampoule is held at that temperature for some time and then cooled. The maximum temperature is usually 20-30°C higher then the melting point of the compound, which makes it possible to obtain the compound in the form of a compact ingot. The rate of heating of the ampoule is governed by the volatility of the nonmetallic component and by the rate of the chemical reaction. If the temperature is not raised sufficiently rapidly, the volatile component may establish a high vapor pressure in the ampoule. If this pressure exceeds the mechanical strength of the ampoule, the latter will explode.

The rate of reaction (and, consequently, the rate of heating) may be increased by vibrational mixing of the charge or melt, i.e., by shaking the ampoule during the synthesis [215].

The pressure in the ampoule may also exceed the safe limit if the compound decomposes (dissociates) at high temperatures. Such a dissociation is accompanied by the evolution of the volatile component whose partial vapor pressure is called the dissociation vapor pressure. If the dissociation pressure of a given compound is higher than the safe pressure in the ampoule, then any attempt to prepare the compound by this method inevitably leads to an explosion. Therefore, the stoichiometric variant of this method is used only to prepare those compounds whose dissociation pressure at the melting point is relatively low.

If the dissociation pressure of the required compound exceeds the safe pressure limit of the ampoule, we can use the nonstoichiometric variant of the direct synthesis at one temperature. In this case, the charge contains some excess of the nonvolatile component. The heating is carried out up to a temperature somewhat greater than the liquidus point, corresponding to a certain alloy composition. The partial vapor pressure of the volatile component in contact with the liquid solution is reduced by the presence of the excess of the other component, and the larger the excess of the nonvolatile component in the charge, the greater is this reduction.

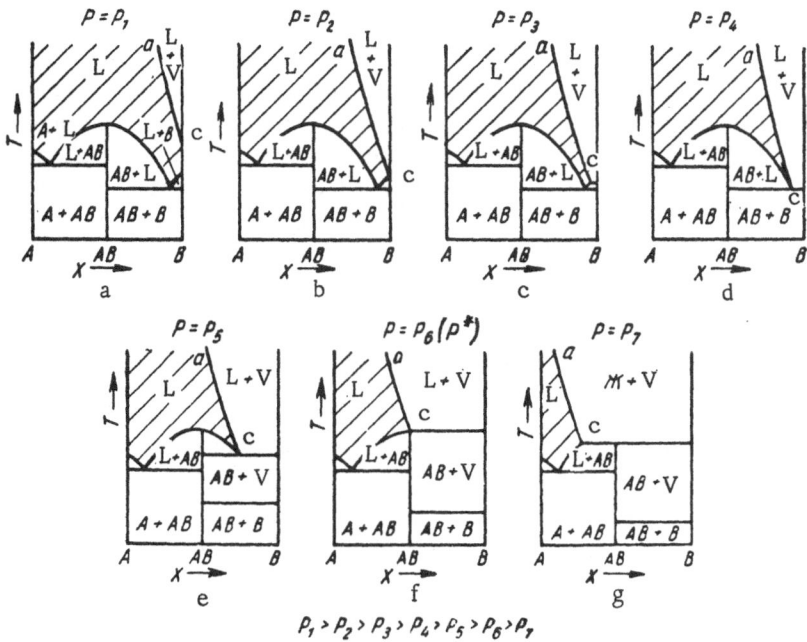

Fig. 4. Isobaric sections through a P−T−X diagram, typical of a binary $A^{III}B^{V}$ system (at various pressures).

The product obtained after cooling the ampoules is a mechanical mixture of the required compound (in the form of smaller or larger crystals or crystalline aggregates) and an excess of the metal component. We can extract the required compound by hot filtering [82] or by selective etching. In the latter case, we need a reagent which will dissolve the excess of the metal but will not act on the required compound.

Direct Synthesis in a Closed Chamber at Two Temperatures

Various modifications of this method are widely used in the synthesis of many semiconducting compounds containing arsenic, phosphorus, and other volatile components. The development of this method followed the determination of the phase diagrams of systems [83] which include semiconducting compounds, showing how the phase composition of a system varies with pressure, temperature, and chemical composition (P−T−X diagrams). Analysis of such diagrams shows that many semiconducting compounds dissociate at high temperatures, giving off a volatile component. Therefore, on heating, we obtain, in general, not the stoichiometric melt of the compound but a solution rich in the metal component and a vapor of the volatile component. In many technological processes we require not a solution but the stoichiometric melt of the required compound.

In order to find a method for preventing the dissociation of a compound, it is necessary to consider the pressure dependence of the phase equilibrium of A^{III}−B^{V}-type systems. This can be done by taking isobaric cuts in the T−X coordinates of the three-dimensional P−T−X diagram. We thus obtain a two-dimensional diagram which is much simpler than the three-dimensional one. However, one must bear in mind that a given T−X diagram applies only at a definite vapor pressure of the volatile component (this applies to any phase diagram of the T−X type, irrespective of how it is obtained).

We shall consider several sections of a P−T−X diagram typical of a $A^{III}B^{V}$ system (Fig. 4). First, we shall deal with a cut obtained at a high vapor pressure of the volatile component which is much higher than the dissociation pressure of the compound at its melting point (Fig. 4a). The characteristic feature of such a diagram, which distinguishes it from the phase diagrams of condensed systems (i.e., systems in which the pressure of the vapor phase is so slight that it can be neglected), is the presence of a region in which the liquid phase and the vapor coexist (L+ V). This region is separated from the liquid state by the curve ca, which indicates how the boiling temperature of the liquid component B varies when the component A is dissolved in it.

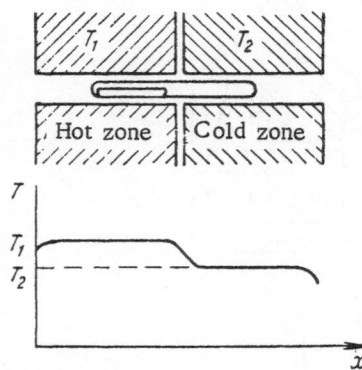

Fig. 5. Schematic representation of a two-temperature furnace and the temperature distribution in two-temperature synthesis.

A section at a pressure P_2 (Fig. 4b) differs from the section just considered in that the L + V region is somewhat greater and the boiling point c of the pure component B coincides with its melting point. Pressure P_2 is the triple-point pressure of the pure component B, i.e., it is the pressure at which a system consisting of only one component B can exist simultaneously in three phases (solid, liquid, and vapor) in equilibrium; this can happen only at a certain definite temperature.

The T−X sections corresponding to lower pressures P_3, P_4, and P_5 show a number of changes in the right-hand part of the phase diagram (Fig. 4c, 4d, 4e).

At a pressure P_6 (Fig. 4f), the line separating the vapor and liquid phases in equilibrium intersects the liquidus line of the system at a point which corresponds to the composition of the required compound. The pressure P_6 is equal to the dissociation pressure of the compound at the melting point.

Any further reduction in pressure leads to a gradual increase of the two-phase region L + V at the expense of the region representing the liquid state. The T−X diagram assumes the form shown in Fig. 4g.

In order to establish the meaning of changes in the nature of the T−X diagram when there is a reduction in the pressure, we shall consider the behavior of samples of the final compound $A^{III}B^V$, heated in an atmosphere of the vapor of the component B at various vapor pressures.

If the compound $A^{III}B^V$ is heated in an atmosphere of the vapor of the component B at a relatively high vapor pressure (P_1), the compound will not dissociate when heated right up to the melting point.

It would seem that, having reached the melting point, we can obtain a liquid solution of stoichiometric composition. However, it follows from the phase diagram (Fig. 4a) that only a liquid of composition given by points lying on the ca line can be in equilibrium with the vapor of the volatile component. Therefore, after the formation a stoichiometric melt of the compound begins to dissolve the B vapor until the composition of of the liquid solution reaches the ca line.

A similar process will occur at lower pressures (P_2, P_3, P_4, P_5). In all these cases, the melting of the compound will be congruent (i.e., the compound will melt without decomposition), but the liquid formed on melting will not be an equilibrium phase. Therefore, at relatively high pressures we cannot obtain a stoichiometric melt: it will always contain an excess of the volatile component.

At the pressure P_6, the melting will still be congruent. The melt formed on fusion may be in equilibrium with the vapor of the component B, since the point on the diagram which represents the stoichiometric melt state lies also on the ca curve. Therefore, the pressure P_6 is the only vapor pressure of the component B at which the stoichiometric melt can exist in equilibrium.

Finally, at lower pressures (P_6, P_7), we cannot obtain a stoichiometric melt because when the sample is heated the compound dissociates, before reaching the melting point, into a liquid of composition c, containing an excess of the metal component and the vapor of the component B. Such behavior of the compound on heating is known as incongruent melting.

Consequently, the pressure P_6 (which will be called P*) represents a boundary above which the compound melts congruently and below which the melting is incongruent.

The following conclusion, important in practice, follows from the above considerations. To obtain a stoichiometric melt of a dissociating compound in a given technological process (synthesis, crystallization, etc.), it is necessary to establish, above the substance, a vapor pressure of the volatile component equal to the dissociation pressure of the compound at its melting point (i.e., the pressure P*).

Thus, if the synthesis of a compound has to be carried out at a temperature higher than its melting point, it is necessary to maintain a vapor pressure P^* of the volatile component in the working chamber. The temperature T^* at which congruent melting occurs is known as the congruent melting temperature of a given compound.

The two-temperature synthesis is carried out in a two-temperature furnace, shown schematically in Fig. 5. The metal component is placed in a boat; the location of the volatile component is of no importance because on heating it is driven from the hot to the cold zone.

Two methods of maintaining the pressure P^* in an ampoule are known. The first requires the regulation of the pressure in the ampoule by varying the temperature T_2 in the cold zone of the furnace. The hot zone is heated to a temperature T_1, somewhat higher than the congruent melting point of the compound. The cold zone is heated to a temperature T_2 at which the saturated vapor pressure of the volatile component P_{sat} is equal to the dissociation pressure of the compound at the melting point, P^*. The process is carried out using an excess of the volatile component so that after synthesis some of this component remains in the cold end of the ampoule. However, in spite of the fact that a charge contains an excess of the volatile component, the resultant ingot is of stoichiometric composition because at the pressure P^* only a stoichiometric liquid can be in equilibrium with the volatile-component vapor at the congruent melting temperature. The duration of treatment, at given temperatures of the hot and cold zones, should be sufficient to allow the initial components to react fully.

In another method of establishing the pressure P^* in an ampoule, one uses a proportion of the volatile component calculated exactly to give the pressure P^* at the end of the synthesis as well as to provide the necessary proportion of the volatile component in the compound being formed. Then the ampoule will not contain a deposit of the volatile component (the vapor will be unsaturated). The hot-zone temperature T_1 is kept, as in the other method, somewhat higher than the congruent melting point. The cold-zone temperature T is established in such a way as to satisfy the condition $T_1 > T > T_2$ (here, T_2 represents the temperature at which the saturated vapor pressure of the volatile component is $P_{sat} = P^*$).

Both these methods apply to the stoichiometric synthesis. However, there are also nonstoichiometric two-temperature methods of synthesis, which involve the establishment, in the ampoule, of a vapor pressure of the volatile component $P < P^*$. These problems are considered in greater detail in [84].

Other methods of direct synthesis are also known but they are quite complicated and that is why they have not found wide application. An example of such a method is synthesis under high pressure (greater than the mechanical strength of the ampoule). In order to avoid the explosion of the ampoule, it is subjected to an external pressure approximately equal to the pressure in the ampoule [85, 86].

The indirect methods of synthesizing semiconducting compounds are very numerous (cf. Chap. II).

Methods of Growing Semiconducting Crystals

Three main groups of methods of growing single crystals are known: from stoichiometric melts, from liquid solutions, and from the gaseous phase.

The most widely used are the methods employing stoichiometric melts. The advantage of this approach is that a fairly simple technique can be used to grow rapidly relatively large crystals. The liquid phase, which is in contact with the growing crystal, has the same composition as the crystal (in the absence of impurities). Therefore, any part of the liquid phase can be used as the supply source of the material for the growing crystal.

In some gaseous-phase methods, the vapor may have the same composition as the crystal. However, the rate of crystallization in this case is limited by the need to transfer a given mass of a substance in unit time from the low-density gaseous phase to the high-density solid phase.

In growing crystals from solutions, the composition of the liquid phase differs from the composition of the growing crystal. In this case, the source of material for the growing crystal is only some of the particles which form the liquid phase. Bearing this in mind and remembering also that diffusion in a liquid is slower than that in a vapor, we may predict that the rate of growth of crystals from solutions is slower than the rates of growth from the melt and from the gaseous phase. Typical rates of growth of crystals are cm/hr from a stoichiometric melt, mm/hr from a gaseous phase, and mm/day from a solution.

The growth of a crystal from a parent phase is an extremely complicated process, which can be regarded as a sequence of a number of simpler processes. Thus, for example, in the growth of crystals from the vapor phase the vapor particles should first "approach" (diffuse to) the crystal surface. Then they should be integrated into a suitable crystal face. This liberates the heat of crystallization, which should be removed from the crystal—vapor boundary to maintain constant temperature conditions at this boundary. Sometimes the process of crystallization is preceded by a chemical reaction in the vapor phase. This increases the number of stages in the process and therefore makes it more complicated.

Thus, the simplest and most rapid method of growing crystals is that utilizing stoichiometric melts. Unfortunately, in many cases this cannot be done. Sometimes the required solid phase cannot be obtained from the stoichiometric melt (for example, when we require the low-temperature modification of a substance which has a number of polymorphic forms). In other instances, the preparation of a stoichiometric melt meets with great technical difficulties. This is usually because of the refractory nature of the substances or due to the dissociation which many semiconducting compounds suffer when heated. In these cases, the preparation of stoichiometric melts is difficult because of the lack of suitable refractory crucibles which do not contaminate the growing substance at high temperatures and the lack of containers which can stand sufficiently high pressures P*.

If there are no suitable materials for the crucibles and chambers in which the chemical reactions are to be carried out, it is convenient to turn to methods of growing crystals from vapor or solution, which can be used at much lower temperatures and vapor pressures of the volatile component.

Main Theorems of the Theory of Crystallization

From the theory of crystallization, it is known that the formation of a crystal from a parent phase proceeds in two stages: the spontaneous nucleation of a three-dimensional crystallization center (crystallization nucleus) in the parent phase, and the growth of this center.

The appearance of three-dimensional nuclei in the parent phase obeys the same general laws which describe spontaneous processes. These laws are considered in detail in chemical thermodynamics (see, for example, [87]) and they reduce to the following main conclusions.

"Since, according to the second law of thermodynamics, the entropy S of a system can only increase (in irreversible processes), ... a system is in stable equilibrium if its entropy is a maximum under given conditions" [87]. Since spontaneous processes proceed in the direction toward the stable equilibrium of a system, the condition for an irreversible process is $\Delta S > 0$.

This relationship can be expressed also in a different way, in terms of a thermodynamic function such as the free energy F (the thermodynamic potential), which is related to the entropy S by a simple expression

$$\Delta F = \Delta U - T \Delta S,$$

where T is the temperature of the system, and U is the internal energy of the system.

Consequently, if the direction of spontaneous processes is governed by the condition $\Delta S > 0$, in terms of the free energy this condition is written as $\Delta F < 0$. In other words, spontaneous processes may proceed only in the direction of decreasing free energy of the system and stable equilibrium systems are characterized by the minimum value of their free energy F under the given conditions.

The free energy of a system depends on the parameters of the state of the system and, in particular, on temperature. Figure 6 shows the temperature dependence of the free energy of liquid and solid phases. At high temperatures (above the melting point of a given substance), the free energy of a system in the liquid state is less than the energy of the system in the solid state. Therefore, in a system at $T > T_{mp}$ consisting of solid and

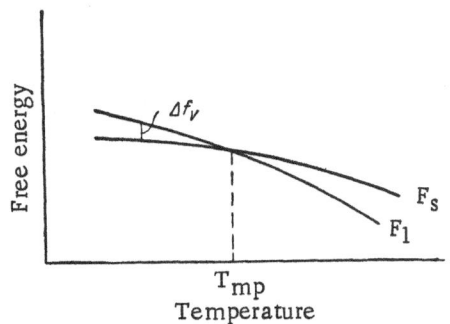

Fig. 6. Temperature dependence of the free energy of a system, F, in liquid or solid state.

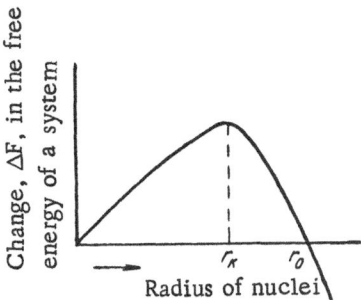

Fig. 7. Dependence of the change in the free energy of a system, ΔF, on the radius of crystallization nuclei formed in it.

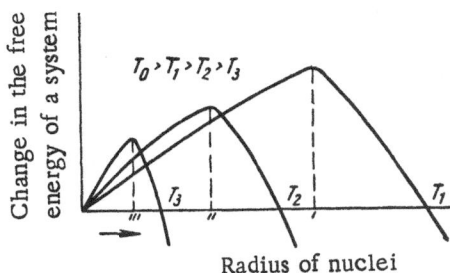

Fig. 8. Change, ΔF, in the free energy of a system with the degree of supercooling.

liquid phases, we can have the spontaneous process of the conversion of a crystal to a liquid (melting) but we cannot have the reverse process whereby a liquid is converted into a crystal (crystallization). Melting is associated with a reduction in the free energy of a system ($\Delta F' < 0$), while the free energy of a system increases on crystallization ($\Delta F' > 0$).

If the temperature of the system is $T < T_{mp}$, the process of crystallization may be spontaneous.

The condition given above is necessary but not sufficient for the process of formation of three-dimensional crystalline nuclei in a parent phase. There is one more factor which governs the changes in the free energy of a system during phase transitions: the formation of a boundary separating the new and parent phases. The formation of this boundary (when a three-dimensional crystalline nucleus appears in the parent phase) requires a certain energy and therefore it always increases the free energy of the system $\Delta F'' > 0$.

Thus, the total change in the free energy of the system when a three-dimensional crystalline nucleus is formed is given by the sum $\Delta F = \Delta F' + \Delta F''$. While the first term $\Delta F'$ is negative when the temperature of the system is $T < T_{mp}$ and positive when $T > T_{mp}$ (when we consider the possibility of the crystallization process), the second term $\Delta F''$ is positive at any temperature.

The values of $\Delta F'$ and $\Delta F''$ (and, consequently, of ΔF) depend on the dimensions of the nuclei being formed. This dependence may have a variety of forms and, therefore, the function $\Delta F = f(r)$, where r is a dimension (for example, the radius) of a three-dimensional nucleus, is not monotonic at $T < T_{mp}$. Such a function is shown graphically in Fig. 7.

The foregoing considerations lead to the conclusion that not all three-dimensional nuclei appearing spontaneously in the parent phase are capable of survival. In fact, if a nucleus of radius $r < r_k$ is formed, its subsequent growth (i.e., the increase in its radius) will increase ΔF. However, as just pointed out, spontaneous processes are possible only if $\Delta F < 0$. Therefore, nuclei of radius $r < r_k$ are not likely to grow but tend to go back to the state of the parent phase, i.e., they tend to dissolve or evaporate. Only those nuclei whose radii $r \geq r_k$ are capable of survival. We can easily see that the growth of such nuclei reduces the ΔF of a system, i.e., their growth is favored by the thermodynamic conditions. The quantity r_k is known as the critical dimension (radius) of a three-dimensional crystalline nucleus. Nuclei of critical dimensions may equally well grow or dissolve (evaporate).

It is evident from Fig. 7 that the appearance of a nucleus capable of survival requires an increase in the free energy of a system (the ordinate $\Delta F > 0$). This energy is provided at the expense of the fluctuation energy. The fluctuation energy required to form a nucleus is maximal at $r = r_k$, but as the radius of a three-dimensional nucleus increases this energy decreases to a vanishing value at $r = r_0$.

The value of the critical radius of a nucleus depends on the degree of supercooling of the melt (supersaturation of the vapor or the solution). At low degrees of supercooling, the value of r_k is large. On increas-

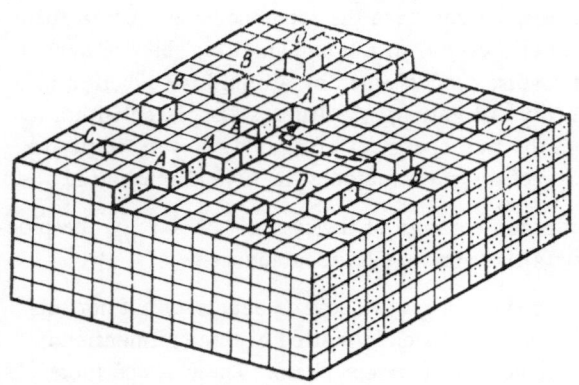

Fig. 9. Schematic representation of a growing crystal surface with an incomplete layer.

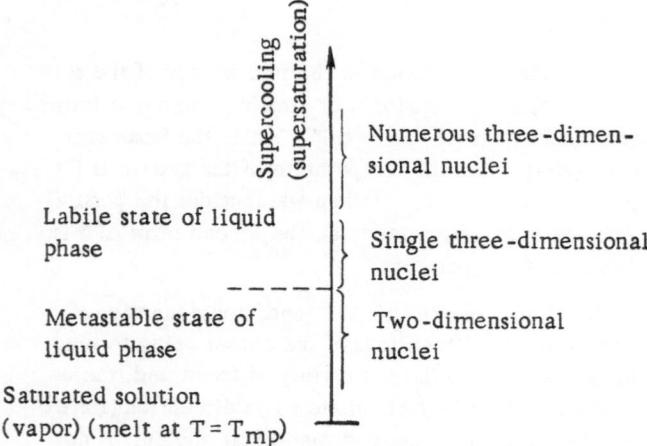

Fig. 10. Qualitative relationship between the degree of supercooling (supersaturation) and the nature of crystallization nuclei.

Fig. 11. Types of container used in growing crystals by the directional crystallization (Bridgman) method.

ing the degree of supercooling, r_k decreases (there is also a simultaneous reduction in the fluctuation energy, ΔF, required for the formation of a nucleus). The qualitative dependence of r_k on the supercooling is shown in Fig. 8.

The probability of the formation of a small three-dimensional nucleus in the parent phase will be higher than the probability of the formation of a large nucleus, especially as the fluctuation energy required to form a small nucleus is less.

Therefore, when the parent phase is severely supercooled and r_k is small, we can expect the appearance of a large number of viable nuclei. This will lead to the growth of a large number of individual crystallites in the parent phase which join up, to produce a polycrystalline solid.

Conversely, to grow a single crystal from the parent phase, it is necessary to establish conditions which would prevent the formation of a large number of viable nuclei in the parent phase. For this purpose, the critical dimension of a nucleus should be large, i.e., the supercooling should be slight.

Thus, the establishment of weak supercooling (supersaturation) makes it possible to limit considerably the number of spontaneous viable three-dimensional nuclei and, in the optimum case, to obtain a single nucleus which, on growing, produces a single crystal.

The second stage of the formation of a crystal is the growth of a viable three-dimensional nucleus after its spontaneous formation in the parent phase.

The fundamental law of thermodynamics referred to earlier is the basis of the theory of growth of a three-dimensional nucleus. Using this law, we can say that a nucleus will grow if its growth reduces the free energy of the system.

The growth process involves the attachment of particles of the crystallizing substance to a three-dimensional nucleus.

It is found that a change in the free energy of a system when a particle becomes attached to a nucleus is governed by the position on the the crystal surface of the nucleus at which this attachment takes place.

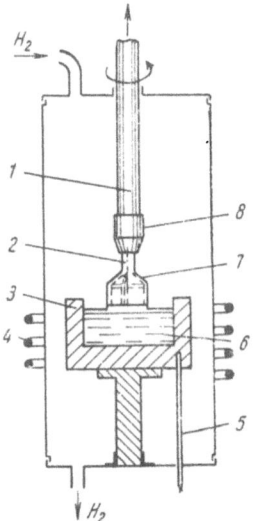

Fig. 12. Schematic representation of apparatus used to pull crystals from the melt: 1) rod; 2) seed; 3) crucible; 4) heater; 5) thermocouple; 6) melt; 7) single crystal being pulled; 8) holder.

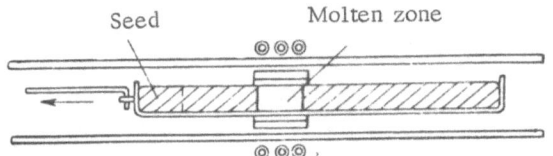

Fig. 13. Schematic representation of apparatus used to grow single crystals by zone melting.

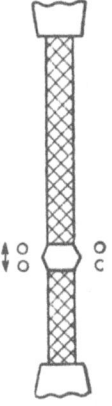

Fig. 14. Floating-zone method.

If we represent each particle by a cube, the crystal surface may be represented as shown in Fig. 9, where an incomplete layer can be seen. The edge of this layer forms an irregular step.

Analysis of the process of crystal growth shows that from the energy point of view the most favorable case is the attachment of a particle to a vacant corner site in the step. Then the volume of the crystal increases but not its surface area (i.e., F' decreases but F" does not increase). The particles reaching a crystal from the melt do not immediately occupy such a corner site: they are adsorbed on the surface, diffuse along the surface to the step (shown by a dashed arrow in Fig. 9), and then move along the step to corner sites where they become attached.

The particles also reach the surface of the crystal when it is not growing but is in equilibrium with the melt. However, then the number of adsorbed particles is equal to the number of particles lost (evaporated) from the surface of the crystal back into parent phase, i.e., we have dynamic equilibrium.

In the case of supercooling (supersaturation), this state of equilibrium is disturbed and the number of particles adsorbed on the crystal surface becomes greater than the number of particles surrendered to the parent phase. The attachment of particles at corner sites produces the motion of a step, i.e., an incomplete layer becomes filled. This process continues until the layer is complete.

Further growth of a crystal requires the formation of an "island" (two-dimensional nucleus), which must appear on top of a completed layer in order to produce new attachment sites.

The spontaneous formation of two-dimensional nuclei obeys the same laws as the formation of three-dimensional nuclei. Therefore, we can speak of a critical dimension of a two-dimensional nucleus, the dependence of this dimension on supercooling (supersaturation), etc. It is necessary to note, however, that for the same supercooling (supersaturation) the critical dimension of a two-dimensional nucleus is less than the critical dimension of a three-dimensional nucleus; moreover, the fluctuation energy needed to produce a nucleus is less in the two-dimensional case. This means that we can have a case when supercooling (supersaturation) is still very weak and three-dimensional nuclei are not formed in the parent phase but two-dimensional nuclei are.

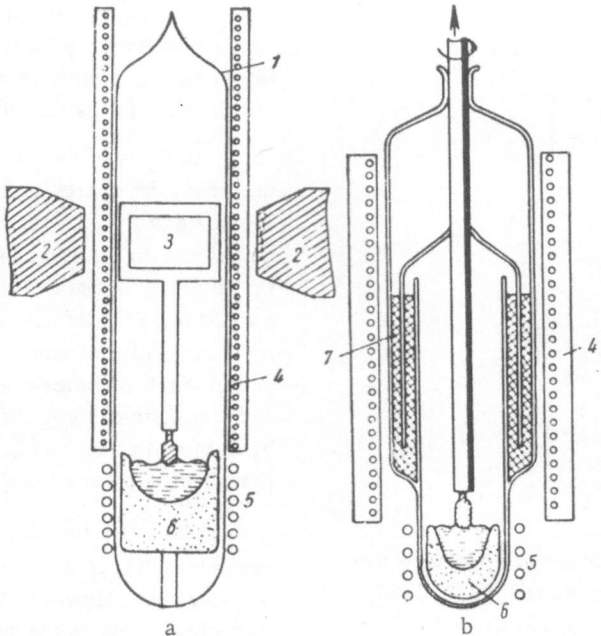

Fig. 15. Apparatus for pulling single crystals of compounds which dissociate at their melting points: a) rod with seed moved by means of a magnet; b) system using a gallium shutter. 1) Hermetically sealed container made of quartz; 2) magnet; 3) alloy with a high Curie point; 4) auxiliary heater which ensures that there are no cold spots at which the volatile component could condense; 5) heater maintaining the material in the crucible in the molten state; 6) crucible; 7) column of liquid gallium.

Figure 10 shows a qualitative relationship between the degree of supercooling (supersaturation) and the nature of the nuclei formed in the parent phase.

This relationship is widely used in the growing of single crystals. We can avoid the initial stage — the formation of a single three-dimensional nucleus (which is very sensitive to supercooling and is therefore quite difficult to control) — by placing a small crystal of the required substance in the parent phase (such a crystal is known as a seed crystal or a seed). In this case, the stage involving the formation of a three-dimensional nucleus becomes unnecessary, since the seed ensures the presence of crystal faces capable of growth. Crystals are then grown using slight supercooling, corresponding to the lower part of Fig. 10. This ensures the growth of a single crystal because the critical radius of the three-dimensional nuclei is then so large that the probability of the formation of such nuclei is negligibly small.

In real crystals, a somewhat different growth mechanism is observed. It is associated with one of the forms of crystal structure defects (dislocations) and is therefore known as the dislocation growth. We shall not describe the latter in detail but merely point out that the characteristic features of the mechanism are associated with the presence of a step, which does not disappear during the growth and which is characteristic of screw dislocations. The presence of such a step dispenses with the stage of the spontaneous formation of two-dimensional nuclei and it therefore makes it possible to grow crystals when there is very slight supercooling (supersaturation).

More detailed information on the theory of crystal growth can be found in the appropriate monographs [88-91].

The Growing of Crystals from Stoichiometric Melts

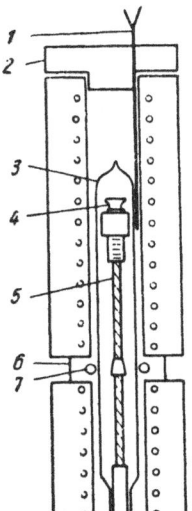

Fig. 16. Hermetically sealed apparatus for growing crystals of dissociating substances using the floating-zone method: 1) thermocouple; 2) stopper; 3) hermetically sealed container made of quartz; 4) rod holder; 5) rod of semiconducting compound; 6) quartz window; 7) high-frequency inductor; 8) auxiliary heater.

One of the difficulties of preparing stoichiometric melts of some semiconducting compounds is the need to maintain the pressure P* in the system, which is equal to the dissociation pressure of the compound at its melting point. If the pressure P* is less than the pressure which a quartz ampoule can withstand, it becomes possible to use those crystal growing methods which have been developed for the preparation of single crystals of nondissociating semiconducting compounds and nonvolatile elemental semiconductors. Among those methods are the directional crystallization techniques, the pulling of crystals from the melt, and zone melting.

The directional crystallization method [92, 93], known as the Bridgman method, involves the cooling of the melt — which is in an ampoule or a boat (Fig. 11) — from one end. The melt becomes supercooled at this end and three-dimensional nuclei are formed in it. The difficulty is to reduce the number of these nuclei to a minimum (in the ideal case, only one nucleus should be formed) and then establish conditions so that only one of these nuclei grows. Further cooling of the melt container is carried out in such a way that the boundary separating the liquid and the growing nucleus travels across the whole liquid along the ampoule or the boat.

The method of pulling from the melt is widely used to prepare germanium and silicon crystals [94-96]. The apparatus used for pulling single crystals is shown in Fig. 12. The melt 6 is in a crucible 3 heated by a high-frequency current or with a resistance heater 4. A seed crystal 2, attached to a rod 1, is placed above the melt; the rod can be moved in the vertical direction. The seed is lowered into the melt where it fuses partly; it is then raised slowly. As the seed is raised, it pulls behind it (due to the action of surface tension forces) a column of liquid which crystallizes as it enters the cooler region above the surface of the melt. The growing crystal has the same structure as the seed so that if the seed is a single crystal, we can pull a single crystal 7 from the melt.

The term zone melting applies to the group of methods in which a narrow molten zone is moved slowly through a relatively long solid sample [97]. The zone-melting technique is widely used and it can be employed to grow single crystals. Figure 13 shows schematically the apparatus used to grow single crystals by this method. The charge consists of a seed, placed at one end of a boat, and the material which we want to convert into a single crystal. This material can be in the form of a number of lumps, a powder, or a polycrystalline ingot. Initially, the heater is placed above the boat in such a way as to melt partly the seed crystal. Then the narrow molten zone is made to move slowly by gradually moving the heater away from the seed. Under certain conditions; this technique can produce a single crystal.

The method in which a crucible is not used and which is known as the floating-zone method (Fig. 14) is also widely used. A molten zone is established in a vertical rod, rigidly fixed at both ends. The liquid does not flow out of the molten zone because of the surface tension. The floating-zone method, like the zone melting in a horizontal crucible, can be used to prepare single-crystal rods. Its advantage is that the molten material is not in contact with the walls of a crucible (in most cases, a crucible is a source of contamination and may give rise to "nonspontaneous" nucleation, crystal structure defects, etc.).

In growing crystals of compounds which dissociate at high temperatures, the need for the growth processes to occur at the vapor pressure P* of the volatile component greatly complicates the techniques of the pulling and floating-zone methods. If a stoichiometric melt is to be in equilibrium with the vapor of the volatile component, whose vapor pressure is P*, the system must not have any "cold spots," i.e., regions where the temperature is lower than that at which the saturation vapor pressure of the pure volatile component is equal to P*. The system should be sealed and should have no metal parts in contact with the reacting vapor of the volatile component. All these requirements are relatively easily satisfied in the Bridgman (directional crystallization) method and in the zone-melting method with a crucible, where the solid phase does not have to be moved relative to the liquid phase.

On the other hand, the apparatus for the pulling of crystals and that for the growing of crystals by the floating-zone method are much more involved and the operation is exceedingly more complex in the case of substances dissociating at high temperatures. The constructional details of such apparatuses are shown in Figs. 15 and 16.

A more detailed description of the apparatus and technique used to grow single crystals of dissociating substances can be found in [98-100].

The Growing of Crystals from Solutions

Depending on the solute content in a solvent, we can distinguish unsaturated, saturated, and supersaturated solutions. Figure 17a shows a typical $T-X$ phase diagram for a $A^{III}-B^V$ system. The region above the liquidus curve represents the liquid state of the system. The regions below the horizontal lines — the solidus lines — represent the solid state of the system. The regions between the liquidus curve and the solidus lines represent the two-phase equilibrium of liquid and solid. Any point M in such a region represents the equilibrium of a solid phase of composition AB in the state K and of a liquid of composition X_L in the state L. The ratio of the liquid and solid phases of these compositions is such that the average composition of the system is X_M.

The point N represents an unsaturated solution, the point O a saturated solution, and the point S a supersaturated solution. The solvent is the component A and the solute is the component B (in considering a quasibinary system $A-AB$, we sometimes consider the compound AB to be the solute).

The most important condition for the growing of crystals is the existence of a supersaturated solution. An unsaturated solution N is thermodynamically stable. At the point O, a mixture of AB crystals and of the liquid of composition X_O is thermodynamically stable. Thus, in these two cases it is not possible to grow crystals. Only the thermodynamically unstable liquid S (a supersaturated solution) can spontaneously decompose into a solid phase AB and a liquid O.

A saturated solution can be made supersaturated by the following two methods:

1. Reducing the temperature (Fig. 17b). A solution C of composition X_C saturated at a temperature T_C becomes supersaturated (C') at a temperature $T_{C'}$. Such a supersaturated solution can decompose spontaneously into crystals of composition AB and a solution C" of composition $X_{C''}$.

Thus, a forced reduction in the temperature of the saturated solution by ΔT leads to the formation of a supersaturated nonequilibrium solution C' and to its decomposition, i.e., to an elementary act of crystallization of the solid phase AB, accompanied by a transition of the liquid to the equilibrium state C".

After such an elementary process, thermodynamic equilibrium is established between the crystals AB and the liquid C".

2. Increasing the concentration of the solution (Fig. 17c). A solution saturated at a temperature T_C becomes supersaturated if, keeping the temperature constant, we increase the concentration of the solution, i.e., if we increase the relative content of the component B. In this case, a forced increase in the concentration of the solution by an amount ΔX leads to the formation of a supersaturated nonequilibrium solution C' which decomposes, i.e., it leads to an elementary act of crystallization of the solid phase AB, accompanied by the transition of the liquid to the equilibrium state C.

After such an elementary crystallization process has taken place thermodynamic equilibrium is established between the crystals AB and the liquid C. The process is isothermal.

In practice we are interested not in elementary crystallization processes but in processes by means of which large amounts of the solid can be obtained. In order to achieve this, it is necessary to establish a supersaturated state in a solution and prevent the liquid from returning from its nonequilibrium state C' to the equi-

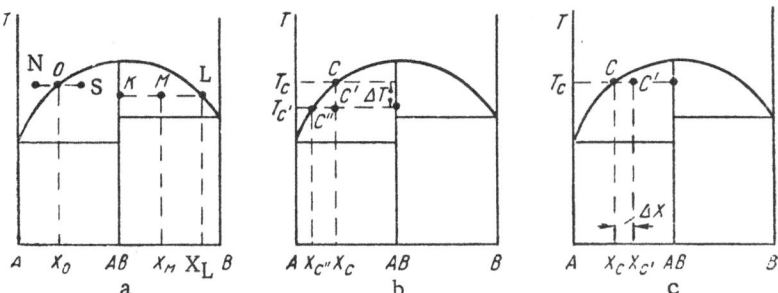

Fig. 17. Diagrams illustrating the methods of preparing a supersaturated solution.

librium state C" (C), i.e., it is necessary to keep the solution in the supersaturated state throughout the whole crystallization process. Then the crystallization will continue as long as the solution is in the supersaturated state and consequently we can obtain large amounts of the crystal phase.

In accordance with the two methods of establishing supersaturation in solutions, all methods of growing single crystals from solutions can be divided into two main groups: those based on the forced cooling of the solution (the composition of the solution is not altered) and those based on a forced change in the solution's composition (the temperature of the solution is not changed and the processes are isothermal).

The temperature of a solution can be lowered in many ways. For example, we can reduce the current supplying the furnace. However, the temperature of a solution may be lowered also without altering the power supplied to the furnace. This is done by establishing a temperature gradient and moving the ampoule in the furnace in the direction of lower temperatures.

The composition of a solution may be altered in two ways. The traditional method (used mainly in aqueous solutions) is to increase the concentration of a solution by evaporating the solvent. The evaporation of the solvent produces a relative increase in the solute content in the solution. However, we can also alter the concentration of a solution (in particular, we can increase it) by altering the absolute content of 'the solute in the solution during crystallization by the process known as forced feeding. The forced feeding of a liquid phase can be achieved by various methods. Some of them are considered below.

There are thus many methods of establishing and maintaining the supersaturated state in a solution. By combining these methods, we can derive a large number of techniques for the growing of single crystals.

Moreover, apart from the cases considered above when the pressure was assumed to be constant, there are crystallization processes which take place at a constant temperature or a constant composition by varying the pressure. In the former case (isothermal conditions), the processes are governed by the parameters P and X, while in the latter (constant composition) they are governed by the parameters P and T. Thus, we can have an isothermal process in which the supersaturation of the solution is achieved by forced changes in pressure, etc. However, such processes have not yet been used to grow single crystals.

The conditions set out above are essential for obtaining a solid phase from a liquid solution. However, to obtain solid phase in the form of a single crystal, a number of additional conditions must be fulfilled.

We have just mentioned that in the growing of crystals it is very important to control the degree of supersaturation by selecting its value so that it is not so high that numerous three-dimensional nuclei are formed and not so low that the probability of the formation of a three-dimensional nucleus is too low. The establishment and maintenance of the optimum supersaturation is one of the main difficulties in the work on crystal growth from solutions.

A widely used technique is to establish supercooling (supersaturation) not throughout the whole volume of the liquid phase but only in a small part of it. By establishing the optimum supercooling (supersaturation) in a small region and keeping the remaining part of the liquid phase at a higher temperature, we increase considerably the probability of the formation of a single three-dimensional nucleus [the number of crystallization centers formed at a given supercooling (supersaturation) per unit time is proportional to the volume of the super-cooled liquid phase]. Such a technique has been justified by the work on stoichiometric melts. However, the situation in the case of solutions is somewhat different. The point is that for nucleation to occur in a solution the solid particles must congregate in a microvolume in an amount sufficient to form a three-dimensional nucleus. The agglomeration of the solute is due to the diffusion of particles, which is a relatively slow process. Therefore, it is not always possible to establish supersaturation in a small volume and maintain it at the required level for a time necessary to concentrate the solute in this volume.

The difficulties of such "seeding" in a selected volume increase as the volume decreases and with the dilution of the solution. Therefore, the role of seed crystals is much more important in solutions.

When crystals are grown from a solution they may be contaminated with the solvent. This may be because of some solid-state solubility of the solvent in the required compound. If there is even negligibly small solid solubility of this type and the solvent is an electrically active impurity in the compound, the resultant crystal may not satisfy the rigorous requirements of semiconductor technology.

The role of the solvent in the growing of crystals from solutions is extremely important. Therefore, the requirements which a solvent must satisfy are quite rigorous.

These requirements can be reduced to the following main points.

1. The use of a solvent should reduce considerably the crystallization temperature of the required compound at the highest possible concentration in the solution.

This condition determines the amount of a substance which can be obtained in the crystalline form from a given amount of melt and depends on the nature of the liquidus curve separating the solvent and the required substance. The most convenient form of the liquidus curve is that when the system "solvent — required substance" forms a eutectic, degenerate in the direction of the solvent.

2. The selected solvent should have a low saturation vapor pressure at the working temperatures so that its partial pressure above the solution should not be high.

3. The solid-state solubility of the solvent should be so small as not to influence the electrical properties of the crystal obtained. It is desirable that the solvent should not be an electrically active impurity in the crystal being grown.

4. The impurity content of the solvent and the segregation coefficient for the impurity distribution between the solvent and the required substance should be such as not to contaminate the material being produced.

If one of these requirements is not satisfied, it is not possible to grow good-quality semiconducting single crystals from a solution.

Moreover, there are additional requirements with respect to the solvent, which, if not satisfied, may make it difficult to obtain crystals from a solution. For example, it should be easy to separate crystals from the solvent; the solvent should be inert with respect to the usual crucible materials, etc.

In spite of these quite stringent requirements with regard to the solvent and some disadvantages of growing crystals from solutions, in many cases this method gives good results.

Unfortunately, many problems associated with the growing of crystals from solutions are not yet fully resolved. Thus, it is not possible to select from theoretical considerations a suitable solvent for a given substance; the phase diagrams of many binary and ternary systems, which include semiconducting crystals, have not yet been investigated, etc.

We shall consider below a number of typical methods of growing crystals from solutions.

Since the difficulties of controlling the growth of crystals from solutions increase as the concentration of the solution is reduced, the use of methods based on forced cooling of the solution (without forced feeding) can be hardly practicable since they are characterized by a reduction in the concentration of the solution during crystallization. The methods described above can be used to deal with concentrated solutions and this inevitably

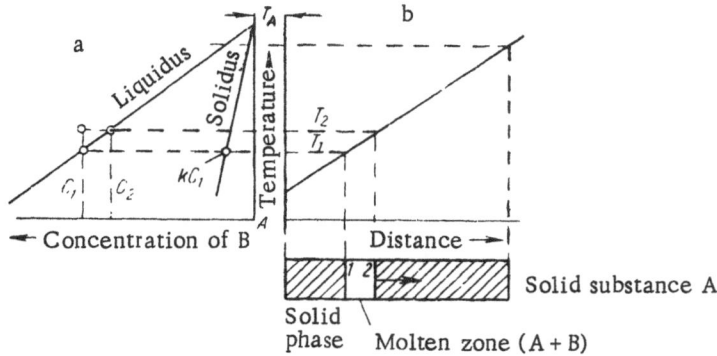

Fig. 18. Melting in a temperature gradient: a) phase diagram of A—B system; b) sample and temperature distribution along its length.

means higher temperatures and pressures, i.e., conditions under which the advantages of growing from solutions are diminished compared with growing from stoichiometric melts.

It is possible to obtain crystals from dilute solutions by slow cooling [101] without the interference of external agencies. However, a large number of crystallization nuclei is then produced and big crystals cannot be obtained. However, the need for large semiconducting single crystals can be disputed because the preparation of a substantial number of semiconducting devices requires only very small single-crystal plates. From this standpoint, the relatively simple method of free crystallization from solutions may be of considerable interest. The much more serious objection is the presence of imperfections in semiconducting crystals grown from solutions by free crystallization. In fact there are indications that the mechanism of growth of diamond-like crystals is basically responsible for such defects as twin planes [337].

To grow larger crystals, it is necessary to use the methods involving the forced feeding of a solution placed in a constant temperature field with a longitudinal temperature gradient. The following methods belong to this group.

Melting in a Temperature Gradient. The technique of melting in a temperature gradient was described by Pfann [97, 102]. Later, it was used to grow germanium single crystals.

Obviously, the same method can be used to prepare crystals or coarse-grained ingots of other semiconducting substances, if a satisfactory solvent can be found. The solute may be in the form of a powder obtained, for example, by indirect synthesis.

The method is based on the difference between the solution — solid equilibrium conditions at the cold 1 and hot 2 boundaries separating the phases (Fig. 18). Solutions of different concentrations are at equilibrium at different isothermal cross sections of the molten zone. Longitudinal concentration gradients produce diffusion currents of the components tending to equalize the composition of the liquid. These currents give rise to supersaturation of the solution and crystallization of the required substance A at the cold boundary 1. At the hot boundary 2, these currents dissolve the substance A and thus produce spontaneous motion of the liquid zone along an ingot in the direction of higher temperatures. The promising nature of this method was excellently illustrated by Broder and Wolff [338], who were able to obtain GaP crystals using gallium as the solvent.

Circulation Methods. This group of methods, widely used at the beginning of the century to grow crystals from aqueous solutions, is based on the same principles as the melting in a temperature gradient [90].

In contrast to the latter method, where the volume of the liquid phase is small, the volume of the liquid phase is large in circulation methods. Therefore, convection may play an important part in the process of equalization of the liquid phase composition, and, consequently, the rate of growth of a crystal may be higher than that in the case of melting in a temperature gradient.

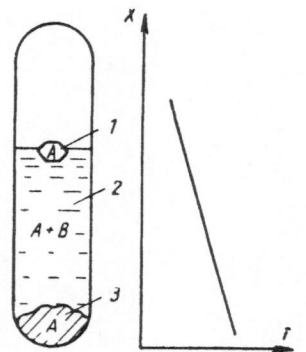

Fig. 19. Circulation method for growing single crystals: 1) crystal of substance A growing in the coldest part of a solution; 2) solution; 3) feed substance A.

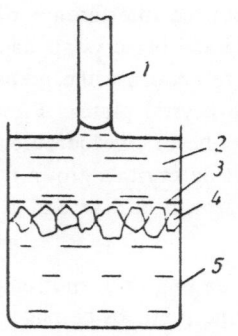

Fig. 20. Schematic representation of the method of pulling a crystal from a melt: 1) pulled crystal; 2) molten solution; 3) quartz net; 4) feed material; 5) crucible.

One of the possible modifications of this method is shown schematically in Fig. 19.

The most serious difficulty in the realization of this method is the problem of separating the substance used to feed the solution and the growing crystal. If the density of a substance A is higher than that of the solution it is necessary to hold, in some way, the growing crystal in the region of the necessary temperatures. If the density of the substance A is less than that of the solution, we have the problem of preventing the feed substance A from floating up to the surface. These problems may be solved by the special construction of the ampoule.

As in many other methods, the use of a seed may be desirable.

The Pulling of Crystals from a Solution. A crystal is pulled from a saturated solution which is in equilibrium with a solid feed material. When the density of the required substance is less than that of the solution, it is necessary to prevent the feed material from floating up to the surface. For this purpose, one can use a quartz net 3 placed in a crucible 5 above the feed material 4 (Fig. 20).

Zone Melting. Zone melting in a crucible can be used also to grow crystals from solutions. However, this method is less convenient than melting in a temperature gradient because in the usual zone melting the molten zone is fed with material of the original composition C_0 which is not the pure required substance. Since the required substance crystallizes after the passage of the zone, the concentration of the solution in the zone will decrease as it moves along. This decrease of the concentration of the solution is undesirable (see above).

The Growing of Crystals from the Gaseous Phase

All methods of growing crystals from the gaseous phase may be divided into two broad groups. One of them includes the methods involving continuous flow systems; the other consists of the methods in which the vapor is in a closed system.

In a continuous flow system, one uses a chemical process (exchange, disproportionation, and thermal dissociation reactions). The initial reacting components (or one component) are fed to a special reaction chamber where the reaction takes place. Under suitable conditions, the required product of this reaction can be obtained in the form of crystals growing on the walls of the reaction chamber or on special substrates. The remaining products of the reaction are removed from the chamber.

Thus, the processes of this type presuppose a continuous flow of the substance through the reaction chamber and this explains their designation as the continuous flow methods.

Such processes can be used to prepare crystals of many semiconducting materials, including refractory substances. The reactions and the reaction chambers used vary widely and are governed by the properties of the compounds to be prepared.

The other group is represented by methods which involve operating in closed systems. The processes in such systems involve the transport of matter in the gaseous phase. Therefore, they can be called gas-transport closed-system processes.

For a gas-transport process to occur, the conditions for thermodynamic equilibrium must be different in different parts of the system. Usually, these differences are achieved by maintaining the various parts of a

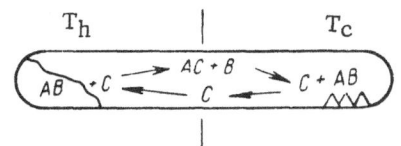

Fig. 21. Gas-transport reaction method.

system at different temperatures. If the equilibrium in the system is purely physical, the gas-transport process involves sublimation or distillation. In this case, the working process is the sublimation of a solid or the evaporation of a liquid. Such processes can be used successfully to grow crystals of those substances whose vapor pressure is not too low at practicable temperatures. Otherwise, the transport of matter is extremely slow. The possibilities of such processes are illustrated very well by the method used by Piper and Polich [339] to prepare large crystals of $A^{II}B^{VI}$ compounds.

When the equilibrium in a system is a chemical equilibrium, the gas-transport process may be defined as the gas-transport reaction. The working process is some chemical reaction. The equilibrium of the initial substances and the final products of such a reaction are governed by the equilibrium constant of the reaction and depend on temperature. By maintaining different temperatures in different parts of the system, we can establish different equilibrium conditions, i.e., different relationships between the amounts of the initial substances and the final products. Since different equilibrium conditions are established within a closed system, the transport of gaseous substances will tend to equalize the composition of the vapor in the working chamber. This transport will disturb the chemical equilibrium. However, in a system which is not in equilibrium, spontaneous chemical processes will tend to return the system to its equilibrium state. This cycle is repeated again and again.

The selection of the working reaction reduces to the selection of a sufficiently volatile substance (transport agent) which would react with a given substance producing volatile products at acceptable temperatures. The reaction equilibrium should not be strongly biased in the forward or reverse direction.

Such a process may take place in an ampoule containing a given substance AB and a small amount of a transport agent C.

Let us assume that the working reaction is an endothermic reaction of the type

$$AB + C \rightleftharpoons AC + B - \overline{Q}.$$

The compound AB and the transport agent C are placed at one end of an ampoule, which is then evacuated and sealed. The ampoule is placed in a two-temperature furnace in such a way that the compound is in a hot zone kept at a temperature T_h (Fig. 21). Under these conditions, the reaction between AB and C continues until equilibrium is established between the initial substances and the reaction products at the temperature T_h. Due to the transport processes, the volatile products of the reaction AC and B enter the cold end of the ampoule, whose temperature is $T_c < T_h$. The reduction in temperature gives rise to the reverse reaction

$$AC + B \rightarrow AB + C + \overline{Q}.$$

At the temperature T_c, the substance AB is in the solid state. Having been formed by the reaction, it precipitates at the ampoule walls. The liberated transport agent C proceeds back to the hot zone and reacts with a new portion of AB; the whole process is then repeated again.

In the example considered, we have the transport of matter from the hot to the cold zone. When the working process is an exothermic reaction, the transport is from the cold to the hot zone.

This method, developed and described by Schäfer [103], produces, in general, polycrystalline material consisting of intergrown crystals. However, under certain conditions, it is possible to deposit a substance in the form of single crystals. For this purpose, it is necessary to control the amount M of a substance AB transported in unit time. The value of M should not be too large in order to allow the growing nuclei to assimilate the incoming material. Otherwise, considerable supersaturation will be established and, consequently, new nuclei may be formed. On the other hand, the value of M should be sufficiently large to grow crystals at an acceptably rapid rate. The value of M for a given transport agent can be regulated by varying the temperature difference between the hot and cold zones, the transport agent concentration, and the ampoule geometry.

Such a method has been used by a number of investigators to prepare single crystals of various semi-conductors, such as chalcogenides [104] and $A^{III}B^V$-type compounds [105, 106, etc.].

The potentialities of this technique are not yet fully developed and therefore the main attention of the investigators of gas-transport reactions has been concentrated not on crystal growing, but on other aspects of these interesting processes, such as transport, purification, etc. Only recently has interest been aroused in the gas-transport reactions as a method of preparing single crystals of technologically complex semiconductors.

The advantages of the method — the possibility of growing crystals of refractory materials at low temperatures, purification during the process, or, conversely, the doping of the transported substance — also include the promising possibility of growing single crystals of solid solutions which are highly uniform in composition. In this respect, the method of gas-transport reactions is without a rival, at least at present.

In conclusion, we must mention the use of gas-transport reactions in epitaxial growth both in continuous-flow systems and in closed systems. In the present book, we shall not consider the methods of preparing epitaxial films, but the technique of epitaxial growth may also be employed to prepare single crystals of technologically complex substances. The substrate, in the form of a single-crystal oriented plate, suitable for the epitaxial growth of a substance, plays the part of a seed crystal. The similarity of the structures and lattice parameters of the substrate and the substance deposited on it are the cause of the growth of a single-crystal layer. The thickness of this layer may become considerable and then the substrate is removed by grinding or etching. Such a technique may be of great interest in the preparation of single crystals of substances which are difficult to grow by other methods (see, for example, [340]).

PREPARATION AND PROPERTIES OF REFRACTORY
SEMICONDUCTING MATERIALS

Diamond

Carbon occurs naturally in three allotropic forms: amorphous (soot) and two crystalline modifications. The two crystalline modifications differ greatly. They are the soft, dark, lustrous graphite and the exceptionally hard, transparent, sparkling diamond.

Of all the substances investigated scientifically in recent years, diamond has probably the longest history [107-111]. Some scientists have found from Hindu religious books that diamond and its unusual hardness have been known in India for more than three thousand years.

The methods of processing diamond were discovered much later. It was found that diamond has a high reflectivity and a high refractive index and that it exhibits strong optical dispersion. At present, the main industrial applications of diamond are based on its exceptional hardness.

Apart from pure, single-crystal diamonds, black crystalline diamond aggregates (carbonado) are found in the natural state.

Carbonado consists of a mass of tightly packed crystallites which have a higher resistance to cleavage under impact and shock than do pure single-crystal samples. Therefore, carbonado is of exceptional interest in technical applications.

In Russia the first samples of diamond were found in 1829 in the western part of the Urals range near the Krestovozdvizhensk Mine. However, the mining of diamond on an industrial scale began only after a geologist, L. I. Popugaeva, found, in 1954, native deposits of diamonds in the Yakut ASSR ("Zarnitsa" pipe). Diamond pipes were later found in outcrops of native diamond-bearing rocks at "Udachnaya", "Mir", "Aikhal" ("Slava") [109]. The Seven-Year Plan for developing the national economy of the USSR for 1959-1965 includes the development of a diamond-mining center in the territory of the Yakut ASSR.

Classification of Natural Diamonds

Diamonds are divided into two classes according to the perfection of the crystals and their external habits. The first class includes large and pure crystals which are used in the jewelry trade, while the second class includes industrial diamonds (bort, ballas, carbonado, etc.) which have defects, are opaque, or are too small to be used as ornaments.

Natural diamonds have a variety of crystal habits. They have well-formed faceting, indicating that they grew in the suspended state, probably from a solution. A detailed investigation of the crystalline habits of natural diamond has been carried out by A. E. Fersman [112]. The most frequently occurring flat-faced types of

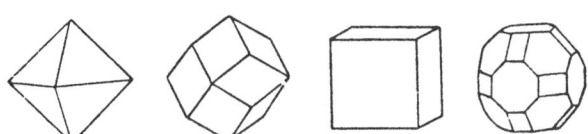

Fig. 22. Usual flat-faced crystal forms of natural diamond.

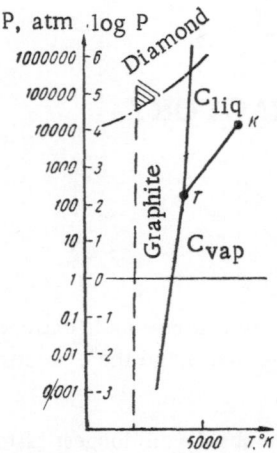

Fig. 23. Phase diagram of carbon.

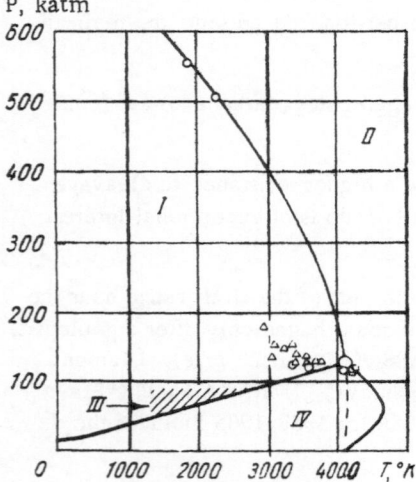

Fig. 24. Part of the phase diagram of carbon, including the triple point diamond—graphite—liquid phase: I) diamond; II) liquid; III) region used to prepare diamonds; IV) graphite + metastable diamond.

diamond are shown in Fig. 22. One often finds spherical-shaped growths and penetration twins on various faces, but the most usual are small crystals (weighing from one-hundredth to one or two carats) although single crystals may weigh as much as hundreds or even thousands of carats (1 carat is equal to 200 mg).

The color of diamonds varies from completely transparent green (which is very rare) and transparent blue (again very rare) to brown and even black.

The jewelry trade prefers the completely colorless or very weakly tinted diamond crystals.

Industrial Applications of Diamonds

One of the main uses of diamonds is in the drilling of wells and holes in rocks where it ensures high efficiency, the long life of drills, and very nearly cylindrical bore holes.

Diamonds are also employed in the accurate drilling of precious stones, hard alloys, watch pinions, etc. Diamond cutters are widely used for machining. Among the other applications are planing, grinding wheels, drawing wire, cutting glass, engraving accurate scales, and ruling diffraction gratings.

Apart from the relatively large diamond crystals, diamond powder is widely used. This consists of very tiny crystals unusable in any other form. Diamond powder is mixed with metal powder and the resultant mixture is pressed in the form of disks which are employed to cut hard materials (including diamond), semiconducting crystals, and stone used for building. Methods have been developed in the USSR and the USA for using synthetic diamonds, mainly in the form of grinding powder and polishing paste.

Recently, diamonds have begun to be used as crystal counters for nuclear radiation [113, 114] and to make some semiconducting devices (see below).

Preparation of Synthetic Diamonds

Attempts to prepare synthetic diamonds date back to the beginning of the nineteenth century. In 1828, Cagniard de la Tour announced that he had prepared synthetic diamond. In 1880, J. B. Hannay reported the preparation of diamond by heating a carbonaceous material together with lithium to red heat in closed iron tubes. In 1920, C. A. Parsons attempted to repeat the experiments of Hannay but was unable to obtain diamonds: the temperature and pressure in Hannay's experiments were far too low. In 1894, H. F. Moissan reported that he was able to obtain fine black diamonds by dissolving carbon in molten iron followed by rapid cooling of the melt in cold water. When the iron was dissolved in an acid, tiny glittering black crystals were obtained. In spite of the successful repetition of these experiments by Crookes in 1909, it has not been reliably established whether these crystals were diamonds, iron carbonates, or spinels

Theoretical calculations carried out in the 1930's of the thermodynamic stability of diamond and graphite have shown that graphite may be transformed into diamond only at extremely high pressures.

Figure 23 shows the phase diagram of carbon, which indicates that the region of the stable existence of diamond lies at pressures greater than 10,000 atm. At lower pressures, the stable form of carbon is graphite.

If the temperature is raised, the minimum pressure needed in the transformation of graphite into diamond increases. Figure 24 shows part of the phase diagram of carbon according to the data of Bundy [115].

The attempts of a number of investigators (mainly Bridgman) to transform graphite into diamond at high pressures have shown that this is impossible at relatively low temperatures. To achieve this transformation, it is necessary to increase the temperature in order to provide the carbon atoms with the energy needed to alter the lattice. However, as was just pointed out, higher pressures are needed at higher temperatures. The dashed line in Fig. 23 shows the minimum temperature at which graphite can be transformed into diamond. This temperature is not less than 1200°C. Bridgman had apparatus by means of which he could establish pressures up to 425,000 atm at room temperature; however, when the temperature was increased to 3000°C, he was able to reach pressures only of the order of 30,000 atm, which were insufficient to transform the graphite into diamond.

The preparation of synthetic diamond has become possible only since apparatus has been available in which high temperatures could be accompanied by extremely high pressures. Such apparatus is described in detail in [116]. Conical steel plungers are used which press against one another to form a small chamber. A cube of the mineral pyrophyllite (hydrated aluminum silicate), whose dimensions are somewhat greater than the dimensions of the chamber, is placed in the chamber. When the plungers are brought together, the pyrophyllite deforms plastically, flows into the gaps between the plungers, and seals the chamber hermetically. Moreover, it serves also as thermal and electrical insulation. In order to subject a sample to high pressure, it is sufficient to drill a hole in the pyrophyllite block (cf. Fig. 39) and to place the material to be tested in it. The sample is heated by passing a current through a metal tube or a helix which is built into the container. The current passes to the heater through the plungers.

To reduce the temperatures and pressures required for the transformation of graphite into diamond, a number of "catalysts" — iron, nickel, manganese, etc. — can be used successfully. The mechanism of the catalytic action is not yet known exactly. We may assume that the transformation of graphite into diamond involves the dissolution of graphite in the catalyst and the subsequent precipitation of the carbon in the form of diamond [116, 117].

Attempts to obtain diamond from graphite at high temperatures without the use of a catalyst gave negative results, even when the pressure was raised to 120,000 atm.

In systems using catalysts, the transformation of graphite into diamond was achieved at temperatures of 1200-2400°C and pressures of 55,000-100,000 atm (this region of pressures and temperatures is shown shaded in Fig 24).

To prevent the resultant diamond from being transformed back into graphite when the system returns to normal conditions, the temperature is lowered to room temperature at the end of the process and only then is the pressure reduced (during cooling, the transformation is "frozen" and the diamond can remain in the metastable state even when pressure is reduced). A number of requirements, which must be satisfied in order to prepare synthetic diamonds, have been given by Bovenkerk et al. [117] and are listed below.

1. The pressure and temperature in the system should be in the region of the stable existence of diamond.

2. The temperature should be sufficiently high for the solution of carbon in the metal catalyst to remain liquid. This requirement determines the minimum values of the pressure and temperature at which the transformation of graphite into diamond is still possible.

3. The further the values of pressure and temperature are from the line representing equilibrium between diamond and graphite (i.e., the further these values are within the region of the stable existence of diamond), the more rapid is the formation of nuclei and the growth of diamond crystals and the smaller their average size.

4. The habits of synthetic diamond crystals depend considerably on the temperature at which they are prepared. At relatively low temperatures, the predominant form is cubic. At higher temperatures, cubo-octahedra and dodecahedra are obtained and, finally, at the maximum temperatures octahedra appear.

The color of synthetic diamonds varies from black (at low transformation temperatures) to green, yellow, and white (at high temperatures). The dimensions of the crystals do not exceed 1-2 mm.

It has not yet been possible to relate the color of diamond crystals to the nature of impurities in them or to the type of catalyst. However, it has been pointed out [118] that the introduction of boron admixtures into the initial charge produces blue or black diamond crystals, while the addition of beryllium or aluminum gives colorless or weakly colored green—yellow crystals.

At 1560°C and 85,000 atm, maintained for 3 min, it has been possible to obtain diamond crystals whose dimensions reached 0.8 mm (nickel was used as the catalyst) [118].

The problem of growing diamond crystals is not solved completely because many questions still remain unanswered. Nevertheless, the method just described is used quite widely, as evidenced by the fact that synthetic diamonds are now produced not only in the USA and the USSR but also in Holland, Sweden, and South Africa.

Properties of Diamonds

Physicochemical Properties. Diamonds crystallize in a cubic lattice forming the characteristic diamond lattice, whose period is 3.56Å. The most important crystallographic forms of natural diamonds are octahedra, rhombododecahedra, and hexahedra (cubes), and also forms with curved and spherical faces. The diamond is the hardest of the known substances *(10 on Mohs' scale of hardness; a microhardness of 8820 ± 1380kg/mm^2 [119], and the hardness of the crystallographic planes differ from one another: hardness(111) > hardness(110) > hardness (100). It is this fact which makes possible the grinding and polishing of diamond with diamond powder.

It should also be mentioned that the hardness of diamonds from different deposits is different. For example, Australian diamonds are much harder than South African.

Diamond crystals are very brittle and have a pronounced tendency to cleave along the (111) planes. Plastic deformation is observed in diamonds only at very high temperatures (1800-2000°C). The elastic constants of diamond, measured by the ultrasonic method in the range from 20 to 200 Mc [326], are $C_{11} = 10.76 \times 10^{12}$ dyn/cm^2; $C_{12} = 1.250 \times 10^{12}$ dyn/cm^2; $C_{44} = 5.758 \times 10^{12}$ dyn/cm^2.

Diamond has a very low thermal expansion coefficient: 4.36×10^{-6} (from 301 to 378°K) and 11.76×10^{6} (from 778 to 878°K) [327]. The Debye temperature, calculated from the elastic constants, is $T_D = 2340$°K.

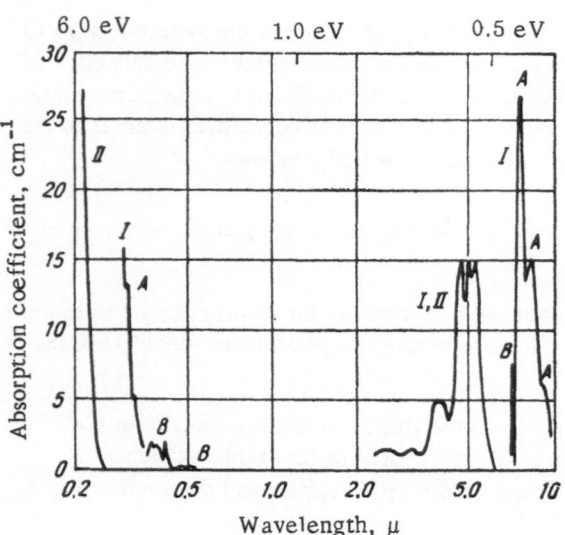

Fig. 25. Absorption spectra of diamond of types I and II in the ultraviolet and infrared regions of the spectrum.

Measurements of the specific heat in the range from 11 to 200°K [120], carried out in a vacuum calorimeter using 8.3 moles (99.992 g) of diamond, gave $T_D = 2240 \pm 5$°K (the two values of T_D given here are extrapolated to T = 0°K).

An unexpected property of diamond is its high thermal conductivity (at room temperature, it is higher than the thermal conductivity of silver). Therefore, diamond, like metals, always seems cold to the touch. This property helps jewelers to distinguish rapidly diamonds from imitations.

With regard to its chemical composition, diamond is pure carbon, but the crystals always contain small amounts of impurities and therefore when diamond is burnt (at 500-800°C in pure oxygen) a certain amount of ash remains: the amount of ash varies from 0.02% for colorless crystals to 5% for less pure samples. The ash usually contains aluminum, silicon, calcium, and magnesium. Diamond crystals may also contain solid (graphite, magnetite, rutile, etc.), liquid (water),

---

*Cubic boron nitride BN exhibits hardness comparable with that of diamond.

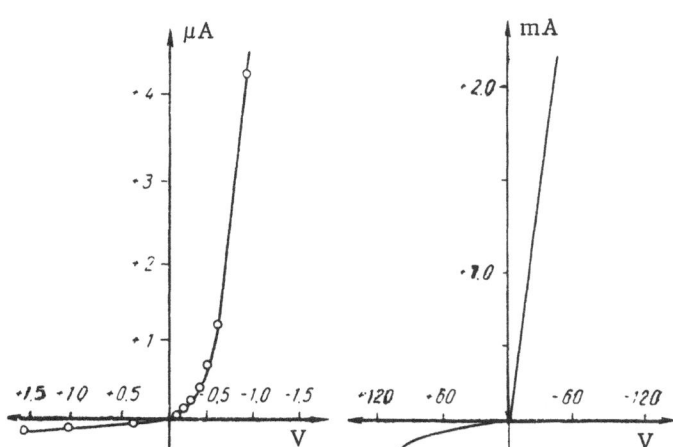

Fig. 26. Rectification at a point contact with a semiconducting
type IIb diamond.

and gaseous (CO_2) occlusions. The specific gravity of diamond depends on the perfection of the crystals and is, on the average, 3.52 (ranging from 3.012 for carbonado to 3.55 for orange-colored crystals).

Diamond is extremely stable under the action of acids and alkalis. It cannot be dissolved at all in hydrofluoric, hydrochloric, sulphuric, and nitric acids, even at high concentrations and high temperatures. Diamond oxidizes slowly and burns in molten sodium and potassium nitrates and in sodium carbonate. On heating in an oxidizing medium to 500-800°C (the temperature depends on the particle dimensions), diamond burns to form CO_2. In air, it is stable to 800°C. Careful investigations, carried out recently [121-123], have established that when a diamond is slowly heated in an oxidizing medium its surface is etched to give triangular etch pits similar to the pits on the faces of natural diamond but with the triangle vertex pointing in the opposite direction. More intensive etching shows blocks and similar structural details [124].

Because of its low atomic weight, diamond is transparent to x-rays and this makes it possible to identify even very good imitations. Diamonds are divided into two types according to their transparency in ultraviolet and infrared light [125]. The usual diamonds are of type I (~ 95% of all diamonds). They have adsorption bands in the infrared region from 3 to 13 μ but in the ultraviolet region they are transparent down to about 3000 Å [126, 127]. In the more rare diamonds of type II, absorption is observed in the region from 3 to 6 μ with a sharp absorption edge near 2250 Å (the forbidden band width is ≈ 5.6 eV).

The absorption spectra of diamonds of types I and II are shown in Fig. 25. It has been established that the additional absorption bands in diamonds of type I are associated with the presence of nitrogen impurity, whose concentration can be as much as 4×10^{20} cm^{-3} [128].

Diamond as a Semiconductor. As established by Custers [129], diamonds of type II can be divided into two subgroups: type IIa and type IIb. The characteristic features of diamonds of type IIb are a strong phosphorescence exhibited after irradiation with ultraviolet light in the 2500 Å range and a relatively low resistivity (50-1200 $\Omega \cdot$ cm), while diamonds of types I and IIa have resistivities of 10^{14}-10^{16} $\Omega \cdot$ cm at room temperature.

Investigations of 21 crystals of natural type IIb diamond having a total weight of 108.00 carats [129] revealed that all these crystals were blue. Their electrical conductivity was measured by determining the current flow through a crystal under a constant voltage of 125 V. It was established that the electrical resistivity varies strongly from sample to sample, reaching a maximum value of 10^8 $\Omega \cdot$ cm. Preliminary experiments showed that the resistivity of blue type IIb diamonds decreased as the temperature rose. Further investigations [130] showed that diamonds of type IIb exhibited rectification when a tungsten probe and a crystal were in contact (Fig.26); the sign of the rectification effect indicated that the investigated samples were p-type. The carrier density (found from measurements of the Hall coefficient) at room temperature was ~ 8×10^{13} cm^{-3} Measurements

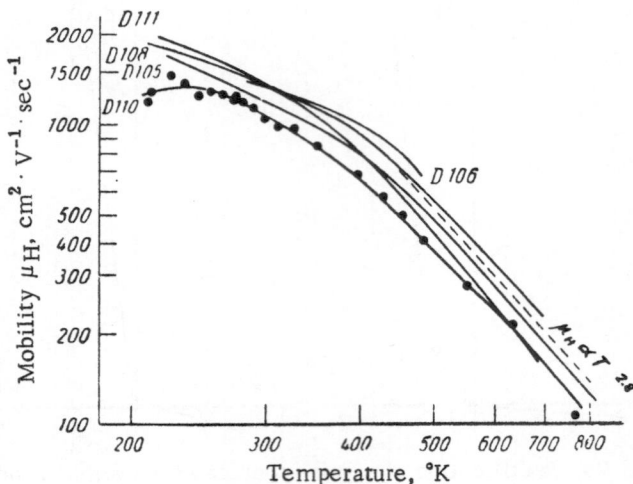

Fig. 27. Temperature dependence of the Hall mobility
for five semiconducting diamonds of type IIb.

of the resistivity and the Hall coefficient of one sample of type IIb diamond [131] in the temperature range from
−100 to + 600°C demonstrated that diamond behaved as a standard p-type semiconductor. The conduction
activity energy was found to be 0.38 eV, and the hole mobility up to 20°C, 1550 ± 150 cm$^2 \cdot$ V$^{-1} \cdot$ sec^{-1}; the
temperature dependence of the mobility was shown to obey the law $T^{-3/2}$. This value of the mobility is in
good agreement with the results of other authors [132-135], obtained for nonconducting diamonds. More detailed investigations of the temperature dependence of the resistivity and the Hall coefficient in the temperature range from 200 to 800°K, carried out on six samples of semiconducting diamond, have shown that at high
temperatures $\mu \sim T^{-2.8}$ [136] (Fig. 27). The experimental data are in good agreement with the theoretical dependences of the carrier density on temperature if $m_p^* = 0.25\, m_0$. The magnetoresistance of diamonds has also
been investigated [136-138] and it has been established that in fields up to 4000 g

$$\frac{\Delta\rho}{\rho_0} = \frac{\rho(H) - \rho(0)}{\rho(0)} = 3.8 \cdot 10^{-17}\, \mu^2 H^2,$$

where $\rho(H)$ is the resistivity in a magnetic field H; $\rho(0)$ is the resistivity in the absence of a magnetic field.

The mobility calculated from the above relationship is 3800 cm$^2 \cdot$ V$^{-1} \cdot$ sec^{-1}, which is obviously subject to experimental error. The linear dependence of $\Delta\rho/\rho_0$ on H^2 indicates that the maximum of the valence
band of diamond lies at the point $\vec{k} = 0$. It has also been established that the longitudinal magnetoresistance
is only one-third of the transverse magnetoresistance. Crystals of semiconducting diamond exhibit phenomena
typical of other semiconductors, such as photoconductivity and photo-emf. The photoconductivity maxima
have been observed at 224, 228, 640, and 890 mμ [139], and the photo-emf appears, apart from these wavelengths, at 440 mμ.

The recently published analysis of the data on the reflectivity of light of energy up to 30 eV [328] makes
it possible to determine some parameters of the energy band structure of diamond by comparing the reflectivity
maxima with certain electron transitions. The observations on the absorption in the infrared region [140] have
shown that semiconducting diamonds of type IIb exhibit characteristic absorption maxima which are in good
agreement with the thermal activation energies obtained from the conductivity or from the thermoluminescence.
It has been established that there are several discrete levels at 0.21, 0.30, 0.37, 0.52, ~ 0.7 eV above the top
of the valence band (Fig. 28).

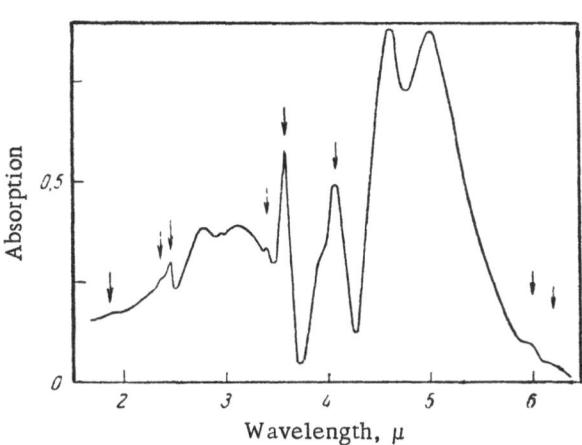

Fig. 28. Absorption spectrum of a type IIb diamond in the
infrared region of the spectrum (arrows indicate the maxima
which occur only in type IIb diamonds).

Investigations of the electrical, optical [141, 142], and photoelectric properties of type IIb diamonds
have firmly established that these diamonds are p-type semiconductors. Wilson [329] has reported measurements of the temperature dependence of the resistivity (in the range 88-293°K) of samples of natural and synthetic diamonds, doped with boron. The resistivities of samples at room temperature ranged from $10^{20} \, \Omega \cdot cm$
for pure crystals to $10^7\text{-}10^{-2} \, \Omega \cdot cm$ for doped ones. Thirty activation energies of the resistivity, ranging from
0.0016 eV to 0.087 eV, were found. The results obtained are ascribed by Wilson to the hopping conduction
mechanism along impurity centers. The main characteristics of semiconducting diamonds are as follows:

ΔE, eV 5.4-5.6

ΔE_A, eV 0.38

m_p^*/m_0 . . . .   0.25

μ_p, $cm^2 \cdot V^{-1} \cdot sec^{-1}$ 1500

μ_n, $cm^2 \cdot V^{-1} \cdot sec^{-1}$ 1800

Temperature dependence of hole

   mobility . . . . $T^{-2.8}$

Photoconductivity maxima, mμ 224, 228,

                                                  640, 890

Applications of Semiconducting Diamonds

In spite of the high cost of semiconducting type IIb diamonds, we can consider the question of their use
in semiconducting devices. This is because diamond is a very convenient material for a number of semiconducting devices which are required to work at temperatures up to 500°C. Keck [164] has reported that, by dif-
fusing aluminum at high temperatures, p-n junctions have been established in diamonds; prototype diamond
transistors have also been made [322].

Semiconducting diamonds are suitable for the preparation of temperature-sensitive resistors (thermistors),
working in the range 200-500°K. The high thermal conductivity, the ability to withstand corrosive chemical
substances at high temperatures, and the low specific heat make it possible to produce strong and reliable
thermistors from blue diamonds.

On the basis of the criteria for the application of semiconducting materials, we may conclude that as
the technique of making synthetic diamonds with controlled properties develops, real opportunities of making
diamond devices will arise.

Silicon Carbide SiC

Silicon carbide SiC occupies a special position among semiconducting materials. It is the only compound of two group IV elements, both of which exhibit semiconducting properties. The properties of this compound are so interesting that SiC has been described as "the bright star on the semiconductor horizon" [143].

The history of SiC is interesting. It was prepared first in 1891 and soon found a number of practical applications (abrasive material, carborundum heating elements, etc.). Because of these applications, technical-grade SiC is now being produced in relatively large quantities (tons). The semiconducting properties of silicon carbide were reported first in 1914, but the preparation of sufficiently large and pure SiC crystals for physical measurements and the construction of devices employing these crystals have proved to be two of the most complex problems in the technology of semiconducting materials.

The preparation of the technical-grade SiC is based on the interaction of quartz sand with carbon according to

$$SiO_2 + 3C = SiC + 2CO - Q.$$

This process takes place in large industrial furnaces with dinas brick lining. Two current electrodes, joined by a coke rod, are placed in the end walls of a furnace. The whole furnace is filled with the charge, consisting of quartz sand and powdered anthracite with an admixture of sodium chloride (added to remove impurities in the form of volatile chlorides). A strong current is passed through the coke rod to heat it and, consequently, the surrounding charge. A layer of crystalline SiC (carborundum), up to 30 cm thick, forms around the rod due to the reaction given above. Crystalline nodes of SiC are formed in the cavities produced at random by the settling of the charge due to the evaporation of the volatile components. The crystals vary in color from white and yellow to grey and black. They contain several percent of impurities (mainly iron, aluminum, and magnesium), which makes them unsuitable for semiconductor applications and forces one to use other methods of preparing SiC in order to obtain the controlled growth of purer crystals. Nevertheless, reports have been published of attempts to determine the physical properties of SiC using the technical-grade crystals [144].

Here, one should mention an interesting property of SiC crystals. It is known that some substances may exist in various crystalline forms — a phenomenon known as polymorphism. Now, polymorphism occurs in SiC crystals, but the various crystalline forms are so similar that a special term "polytypism" has been proposed to denote polymorphism of this type [145, 146]. Only one cubic form of SiC is known; it has the sphalerite structure and is called β-SiC. All the other forms of SiC (there are more than 15 of them) are designed α-SiC.

Methods of Growing Silicon Carbide Crystals

SiC crystals may be obtained in three ways: by sublimation, by growing from liquid solutions, and by growing from the gaseous phase using certain chemical reactions. All these methods are considered below.

The Growing of Silicon Carbide Crystals by Sublimation. This method, used by Lely [147, 148], was the first to be employed in the controlled growth of SiC crystals purer than those of the technical grade. Lely used lumps of technical-grade SiC as the raw material. These lumps were placed in a hollow graphite cylinder, closed at one end, which was located in a tubular graphite heater. The charge was heated to 2500°C in an inert atmosphere. After several hours, SiC crystals were formed on the walls of the cylinder.

The method used by Lely to prepare SiC crystals was a considerable step forward compared with the manufacture of technical-grade SiC. However, this method had several disadvantages, the main one of which was that the raw material was the relatively impure technical-grade silicon carbide. Therefore, crystals grown by sublimation contained considerable amounts of impurities. Moreover, the use of SiC lumps prevented Lely from establishing uniformly varying (in space) temperature conditions [149]. The crystals obtained were mostly hexagonal pyramids: only their basal face was smooth, the other faces had steps. Such crystals are less convenient for physical measurements than plane-parallel plates.

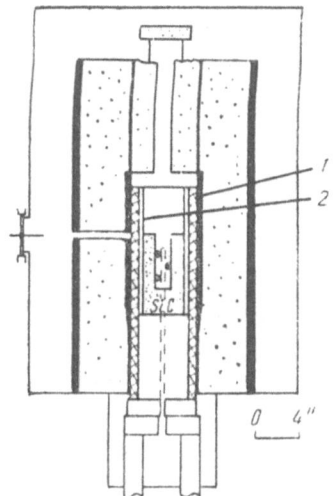

Fig. 29. Furnace for growing SiC crystals by sublimation: 1) cylindrical graphite heater; 2) thin-walled graphite closed cylinder containing SiC.

Attempts have been made to improve Lely's method [149, 150, 151].

We shall now consider the process of growing SiC crystals by sublimation, as described in Hamilton's papers [149, 151]. Figure 29 shows a section through a Hamilton furnace. It consisted of a cylindrical graphite heater, insulated from a stainless steel container by lampblack. The container was placed in a water-cooled vacuum reaction chamber. The furnace could work at pressures down to 10^{-5} mm Hg and at temperatures up to 2000°C. At higher temperatures, the materials of which the furnace was constructed began to vaporize.

In an atmosphere of hydrogen or argon, the furnace could work at temperatures up to 2700°C.

A thin-walled graphite cylinder, closed at one end, was placed inside the graphite heater. SiC, or a mixture of silicon and carbon, was placed in this cylinder in such a way as to leave a cylindrical cavity (well) in the middle.

The charge prepared in this way was heated to a high temperature. The charge was vaporized and the vapor moved from the hotter parts of the furnace along the well to the colder parts. There, the vapor was found to be supersaturated and plate-like crystals could grow from it.

The working temperature was varied from 2290 to 2700°C. It was found that the rate of growth of crystals increased considerably at higher temperatures and that the majority of the crystals grown at higher temperatures were plate-like and therefore suitable for measurements and the preparation of devices. However, increasing the working temperature roughened the crystal faces. These rough faces were in sharp contrast to the smooth faces obtained at lower temperatures. Moreover, the structure of the crystals grown at the higher temperatures was much more imperfect (the crystals had high dislocation densities).

At the temperatures cited above, α-SiC crystals were obtained. Below 2200°C, cubic β-SiC crystals predominated The phenomenon of epitaxial growth (i.e., the regular intergrowth of crystals of different substances or different modifications of the same substance) was observed in many cases: β-SiC crystals grew on the basal faces of hexagonal silicon carbide plates, which had been grown at higher temperatures. Such growth occurred on cooling the furnace after a run.

These experiments were carried out in an argon atmosphere. Sublimated SiC can be obtained also in a hydrogen atmosphere, but the required growth conditions are more difficult to maintain. Moreover, the yield of plate-like α-SiC crystals is lower in hydrogen.

The growth of hexagonal SiC "whiskers" (based on screw dislocations) was observed in a hydrogen atmosphere. The diameter of these whiskers was 5-50 μ and their length 10-20 mm.

It was found that the purity of grown SiC crystals was affected considerably not only by the purity of the original SiC and by the absence of contamination in the furnace itself, but also by the nature of the atmosphere used and by the working temperature. Thus, it was found that SiC crystals grown in hydrogen contained less nitrogen (donor) than those grown in argon (for the same concentration of nitrogen in both atmospheres and other conditions being equal). This may be due to the difference between the diffusion coefficients of nitrogen in hydrogen and argon and due to other features of the transport processes in these media. On the other hand, the content of nitrogen in SiC crystals grown in an argon atmosphere at low temperatures was considerably higher than in crystals grown at high temperatures. This can explain the appearance of p-n junctions in some crystals of technical-grade silicon carbide [144].

The technique of growing crystals by sublimation is quite difficult because it requires special high-temperature furnaces, as well as the establishment and maintenance of exact temperature conditions. This is mainly

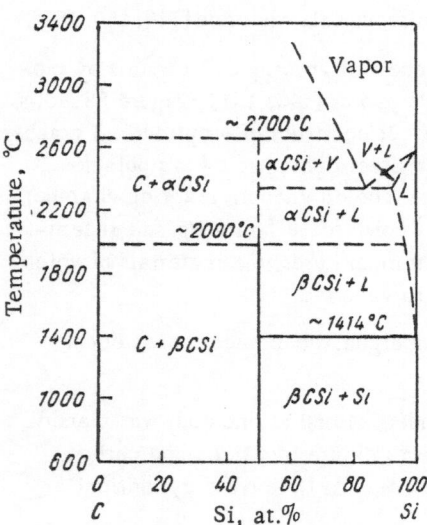

Fig. 30. Phase diagram of the C—Si
system at P = 1 atm.

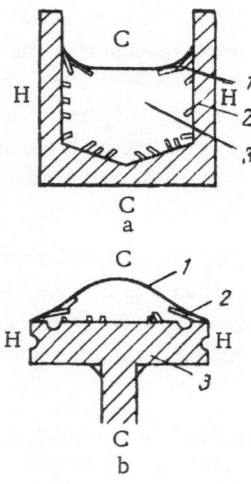

Fig. 31. Methods of growing SiC crystals
from solution: a) in a crucible: 1) SiC;
2) crucible, 3) solution, b) using the
"pedestal" technique: 1) solution, 2)
SiC, 3) pedestal.

because the solid SiC is in equilibrium not with the stoichiometric vapor (i.e., vapor of the same composition as SiC), but with a vapor considerably enriched with silicon [152].

The dimensions of plate-like crystals, obtained by the method described above, can be as much as several millimeters (up to 10 mm) and even larger.

The Growing of Silicon Carbide Crystals

from Liquid Solutions. As mentioned in the first part of the book, the utilization of crystallization from liquid solutions makes it possible to reduce considerably the working temperature. This is extremely important for SiC and therefore many attempts have been made to grow it from solutions. The reduction in the working temperature should improve the crystal structure (for example, reduce the dislocation density) and simplify the apparatus and control procedure. Moreover, one can use seed crystals in the growing of crystals from condensed media.

However, the preparation of SiC crystals from liquid solutions raises a number of problems. These are associated with the difficulty of finding a suitable solvent for silicon carbide and with the selection of a crucible material which could contain the molten solution.

Experiments have been described on the use of elements, other than silicon and carbon, as solvents. For example, iron and nickel are among solvents that have been used. However, since there is a danger of the formation of solid solutions of these metals in SiC, it is more desirable (from the point of view of the purity of the grown material) to use one of the constituents of silicon carbide, namely, silicon, as the solvent. But, it is then very difficult to select a crucible material which would not interact with silicon, the latter having an exceptionally high chemical activity at elevated temperatures. If graphite is used as the crucible material, the system — consisting of a growing crystal, solution, and the crucible — has only two constituents: silicon and carbon.

Ellis [153] melted silicon in a graphite crucible, maintaining the required temperature by means of a graphite resistance furnace or using induction heating. Carbon dissolves in liquid silicon (Fig. 30) and its solubility increases with temperature [154, 155]. Thus, when the melting point of silicon is reached and as the temperature is further raised, the carbon of the crucible is dissolved in the melt. The concentration of the saturated solution formed in this way is governed by the appropriate liquidus line

Since all points in the crucible are not at the same temperature, we can expect that at certain relatively cold spots the solution will be supersaturated and solid SiC will be precipitated from it. This depletes the carbon content of the melt and allows the melt to dissolve the next portion of carbon. This process is none other than the circulation method described in the first part of the book. The characteristic feature of this modification of the circulation method is that the feed material is the crucible whose walls become thinner as the crystallization progresses.

This process has been used in two variants. In one, silicon is placed in a graphite crucible, while in the other it is placed on a graphite disk 20 mm in diameter (Fig. 31). Crystals are formed in the coldest parts of the melt and they float toward the solid phase, where they become attached. Further growth is determined by the direction of heat transfer. The letters H and C denote the hot and cold parts of the systems.

Using this method at 1600°C, it has been possible to grow crystals in the form of irregular prisms whose height reached 5 mm after 3 hours at the maximum temperature. The crystals were yellow β-SiC.

To accelerate the growth of SiC crystals, it is necessary to increase the concentration of carbon in the solution, i.e., to increase the temperature. However, an increase in temperature to 1950°C is accompanied by "filtration" of the liquid solution through the crucible walls. The solution seeps through the graphite, reaches the graphite heater, and soon puts it out of action. The seepage of liquid silicon through graphite has been observed also at lower temperatures — for example, at the melting point of silicon (1410°C) — in those cases when the density of graphite has been below 1.75 g/cm^3. However, the use of dense types of graphite has made it possible to eliminate this effect. At higher temperatures (up to 1950°C), the use of even the densest types of graphite does not prevent seepage.

Ellis reported extremely interesting data on the influence of different amounts of the same impurity on the nature of the crystal growth and on the crystal properties. The impurities used were Li, Mg, B, Al, Ga, Ge, As, Sb, Cu, Fe, Ni, and Au, and the amounts of each of these introduced were 0.5, 0.05, 0.005, and 0.0005 g. The crystallization was carried out in a graphite crucible for 2 hours at 1600°C. Five grams of silicon were placed in the crucible.

It was found that as the impurity content was increased the total yield and dimensions of the crystals decreased, the crystals became more imperfect, the coloring became more complex, and the resistivity decreased.

In addition the following observations were made.

1. An excess of silicon in SiC led to the appearance of n-type conduction. Admixtures of Al and B did not produce p-type crystals if their content in the solution was less than 1-5%.

2. An admixture of copper produced p-type crystals.

3. All the other impurities either gave rise to n-type conduction, or did not enter the crystal lattice in amounts sufficient to compensate the excess of silicon.

Halden [156] also used a graphite crucible, heated by a graphite resistance heater. Halden employed the same circulation method as Ellis. However, Halden grew SiC crystals using a seed supported by a holder and immersed in the coldest part of the melt. The holder could be raised or lowered by an electric motor at rates varying from 0.006 to 0.3 mm/min. The crucible could also be lowered or raised. Moreover, the seed holder and crucible could be rotated. The heater produced a zone ~ 6 cm long with a uniform temperature distribution and a zone above it with a temperature gradient of the order of 27 deg/cm. The hot zone of the furnace was surrounded by molybdenum screens, which reduced the heat lost by radiation, and a double-walled jacket through which cooling water was passed.

The furnace was heated in vacuum of the order of 10^{-4} mm Hg. Then, an inert gas was admitted into the furnace and the temperature was raised to the working value. The crucible was held for half an hour in the uniform-temperature zone, then it was raised into the zone with a temperature gradient and a seed was lowered into it. In a typical case, the temperature at the bottom of the melt was 1750°C. Seed temperature was 1665°C.

Because of the unavailability of SiC crystals, Halden used first a graphite needle as a seed. On being immersed in the melt, the needle became saturated with silicon and interacted chemically with it. During this process, the graphite needle fractured and lost its initial shape so that the end of the needle described a circle of 2.5 cm diameter when the rod to which it was attached was rotated. When it became obvious that the seed had begun to grow, it was raised at a rate of 0.01 mm/min. This rate was slowly increased, reaching finally 0.06 mm/min. After 9 hours, a crystal ingot of up to 25 mm in length and 6 mm in diameter was pulled from the melt. The excess of silicon was etched away leaving an aggregate of small transparent yellow crystallites, which were easily separated.

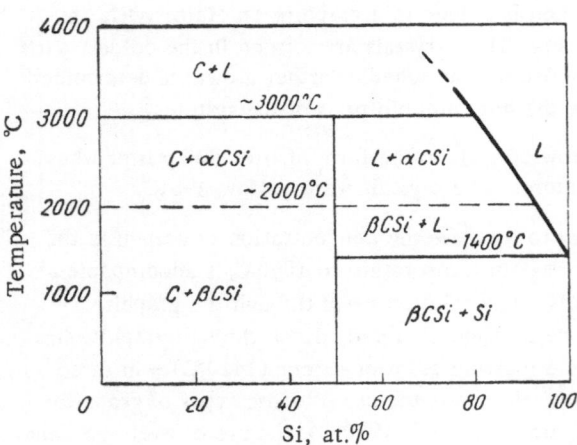

Fig. 32. Phase diagram of the C−Si system at P > 300 atm.

Simultaneously with the crystallization of SiC at the seed, crystallization occurred at the crucible − melt boundary, where about 2 g of crystals grew. The shapes of the crystals were very varied: needles, dendrites, plates, and cubes. Single-crystal needles reached 4 mm in length and 0.4 mm in diameter.

In later experiments, needle-shaped silicon carbide crystals were used as seeds. Also, the longitudinal temperature gradient was decreased (by placing, above the crucible, a graphite screen which had only a small aperture for the seed and a narrow slit for making the temperature measurements). It was found that the decrease in the longitudinal temperature gradient reduced the rate of growth and improved somewhat the structure of the crystals.

When the temperature of the melt was reduced so that its lower regions were at 1525°C and the upper ones at 1475°C, it was found that the quality of pulled crystals improved considerably. Initially a crystalline "raft" grew at the seed, which was followed by the growth of needles and plates. The mixing of the melt was found to be very important: if the crucible or the seed was not rotated only the "raft" was formed [153-155].

The experiments described here represent only the first attempts at the controlled growth of SiC crystals from solutions. There are still many problems to solve and many questions to answer. One of the most important problems is the crucible material. The idea of using an SiC crucible is extremely attractive. However, at present such a crucible can be made only of technical-grade SiC, which is insufficiently pure. Another problem is associated with the fact that even if a single-crystal seed is used, uncontrolled nucleation cannot be avoided. This difficulty will probably be solved by a more careful control of the temperature conditions during growth, and by increasing the amount of carbon in the melt. However, the use of more concentrated melts unavoidably introduces new problems. In order to increase the concentration of carbon in the melt, it is necessary to raise the melt temperature, and this increases the partial pressure of silicon vapor, which is considerable even at 1700°C (at 1880°C the saturated vapor pressure of pure silicon reaches 10 mm Hg [157]). Thus, the growing of SiC crystals from a silicon solution contained in a graphite crucible is a process in which the solvent evaporates quite rapidly and the crucible is dissolved in the liquid filling it. These circumstances make it very difficult to control the crystal growth.

The considerable rate of evaporation of silicon is obviously a serious difficulty in the use of melt methods which dispense with the crucible, but these methods are extremely interesting in connection with the growing of SiC crystals.

The region of liquid solution of SiC in silicon is relatively small if the system is at 1 atm (cf. Fig. 30). However, with an increase in pressure, the region of the liquid state in the silicon−carbon system becomes considerably larger. According to Scase and Slack [155], the solubility of carbon in liquid silicon at 35 atm reaches 19 at.% at 2830°C. At pressures greater than 300 atm, the phase diagram may assume the form shown in Fig. 32. Thus, one of the possible ways of obtaining more concentrated solutions is to grow SiC crystals under pressure. However, this introduces further difficulties.

The use of metals as solvents for SiC is linked to the desire to avoid the difficulties in working with silicon at high temperatures and with the attempt to find a liquid in which more SiC could be dissolved than in silicon and at relatively low temperatures. In this connection, it is worth mentioning Halden's report [156] that the solubility of carbon in a solution of iron and silicon (70% Fe + 30% Si) is 8-10 times greater than the solubility of carbon in liquid silicon (at the same temperatures within the range 1500-1700°C).

Ellis [153] reported the preparation of very small SiC crystals from solutions containing iron and silicon or nickel and silicon (silicon was added to prevent the precipitation of graphite from the solution). The process involved the recrystallization of SiC crystals from a solution. A charge consisted of 1 g of yellow β-SiC crystals, 100 g iron, and 5 g silicon. This mixture was heated in a quartz crucible until a homogeneous liquid was produced and then the melt was cooled at the rate of 100 deg/hr until it solidified. The SiC crystals were separated from the solvent by treatment with a solution of nitric and hydrofluoric acids. The crystals obtained were small (up to 0.6 mm), which was not unexpected for such a rapid cooling. The crystals were green, exhibited n-type conduction, and, in contrast to crystals grown from pure silicon, were α-SiC. Hence, Ellis concluded that the α-SiC form was "stabilized" by impurities.

The growing of SiC crystals from a melt consisting of 65% iron and 35% silicon is mentioned by Halden [156]. The melt was kept at 1650°C. Crystals grew at the crucible — melt boundary in the form of thin hexagonal transparent plates, whose color ranged from yellow to green or blue.

Very few experiments have been carried out so far on the growth of SiC crystals from solutions using metal solvents. The main reason for the caution in using such solvents is the possibility that the growing SiC crystals may be contaminated with the solvent. Therefore, silicon is preferred as the solvent. However, bearing in mind the difficulties in working with solutions based on silicon and the fact that an excess of silicon may enter the SiC lattice and cause n-type conduction, one should turn more to the use of metal solvents and at least study in detail the possibilities of their use.

The Growing of Silicon Carbide Crystals from the Gaseous Phase Using

Chemical Reactions. We shall now consider the methods of preparing silicon carbide which are based on some chemical reaction taking place in the gaseous phase. Such reactions have been used to prepare various coatings and high-purity metals as far back as the end of the last century. Thus, Lodygin was successful in depositing a layer of tungsten on a carbon spiral by heating the spiral in a mixture of hydrogen and tungsten hexachloride vapor. In the first half of the present century, this method was applied successfully to the preparation of tantalum, titanium, zirconium, and other refractory metals in pure form. Since 1935, similar techniques have been used to prepare protective coatings with special properties [158]. Finally, the first silicon detectors for radar were prepared using a similar method: a layer of silicon, formed by thermal dissociation of $SiCl_4$, was deposited directly on a crystal holder.

The method is based on the reaction between the initial substances at a definite temperature. The initial substances should be sufficiently volatile compounds (usually halides). The apparatus used in such processes consists of two main parts: an evaporator and a reaction chamber. The evaporator contains the initial substances and provides the vapor. The vapor is transferred, by a transport gas (or without such a gas), at a controlled rate to the reaction chamber. The chamber (or some part of it) is heated to a temperature at which the required reaction proceeds at a sufficiently rapid rate. The required substance is produced as the result of this reaction and, under suitable conditions, the reaction product is deposited on the reaction-chamber walls or on special substrates (in the case of coating processes, the product is deposited on the objects to be coated). If the temperature at which the reaction takes place is lower than the melting point of the required substance, the latter is deposited in crystalline form. Under suitable conditions, the substance can be obtained also in the form of single crystals. The chemical reactions used in such processes may be reduction or substitution reactions, or thermal dissociation processes.

This technique has found wide application in the metallurgy of semiconductors, in particular, in the preparation of pure silicon. There are published reports of methods for preparing silicon by reducing it from $SiCl_4$ [159], and $SiHCl_3$ [160], and by the thermal dissociation of such compounds as SiI_4 [161, 162], SiH_4 [163], etc.

The advantage of the method lies mainly in the fact that initial substances having low boiling and melting points can be easily obtained in a very pure state by the application of classical physicochemical purification methods. On the other hand, the required substance, in particular, single crystals of it, may form in the gaseous phase at temperatures considerably below its melting point.

It was for these reasons that some investigators, who have attempted to prepare crystals of pure silicon carbide, have turned to this method. The following initial substances have been used with success: 1) $SiCl_4$ + + a mixture of toluene $C_6H_5CH_3$ and hydrogen; 2) trimethylchlorosilane $(CH_3)_3SiCl$ + argon (used as the transport gas); 3) methyltrichlorosilane CH_3SiCl_3 + hydrogen; 4) methyltrichlorosilane + toluene + hydrogen; 5) silicon vapor + a mixture of toluene and hydrogen (argon) [164].

These mixtures were placed in a reaction chamber which was either heated directly or had a carbon heating filament. Experiments showed that a system with a heating filament, used successfully, for example, to prepare silicon by the iodide method [161] could not produce relatively large silicon carbide crystals. The growth of crystals on such a filament stopped when, due to the cooling of the crystal surface, they reached dimensions of 0.1 mm. Increasing the filament temperature did not improve the results because crystals began to be sublimated at the contact with the filament and dropped away from the filament. Therefore, in the majority of cases, the crystals were grown on a hot wall of the graphite reaction chamber.

Various constructions of the reaction chamber were used. Sometimes the chamber was a graphite crucible, into which a graphite tube was lowered, and the reacting gaseous mixture was supplied along this tube from the evaporator. In other cases the reaction chamber was in the form of a graphite tube through which the reacting mixture was passed.

The initial substances in the evaporator were kept at strictly fixed temperatures. For this purpose, liquid thermostats or dewar vessels were used and flasks containing the initial substances were placed in them.

The transport gases were carefully purified before being used in order to remove oxygen, moisture, and organic impurities.

The parameters which governed the growth of the crystals were the reaction temperature, the nature of the transport gas (reducing or inert), the nature of the carbon-bearing and silicon-bearing components, the molar ratio of the two reagents in the gaseous form, the total concentration of the reagents, and the rate of supply of the gaseous mixture to, or through, the reactor. All these parameters had to be carefully controlled in order to obtain crystals of optimal dimensions and composition [159-163].

The optimal combinations of the parameters were found for each initial gaseous mixture. We shall consider Brenner's work [165] which is the most typical of the gaseous methods. Brenner investigated the possibility of preparing silicon carbide crystals by the dissociation of methyltrichlorosilane CH_3SiCl_3 in a hydrogen atmosphere using a graphite crucible which was heated by high-frequency currents. The initial mixture consisted of methyltrichlorosilane, purified by fractional distillation, and hydrogen. The fractional disillation made it possible to reduce the impurity content in the methyltrichlorosilane from 5% to less than 10^{-4}%. Electrolytic hydrogen of 99.9% purity was passed through a drying column and through a counter which controlled the rate of flow of the hydrogen. Next, the hydrogen was passed through a chamber containing methyltrichlorosilane, placed in a thermostatted bath. Here, the hydrogen was saturated with methyltrichlorosilane vapor and passed through a mixing chamber, filled with glass wool, to a graphite tube inserted in a graphite crucible.

The gaseous phase composition was controlled by varying the rate of flow of the hydrogen and the temperature of the methyltrichlorosilane.

The crucible (made of spectroscopically pure graphite) was 15 cm high and ~4 cm in diameter; its walls were 10 mm thick.

The crucible was insulated thermally by magnesium oxide, by a quartz tube, and by mica. The temperature was measured with an optical pyrometer.

The duration of the process ranged from 1 to 8 hours; the hydrogen was admitted at the rate of 20-130 cm^3/min; the concentration of methyltrichlorosilane ranged from $2 \cdot 10^{-4}$ to $4 \cdot 10^{-3}$ mole/min; the crucible temperature was varied from 1200 to 2200°C.

The experiments established that temperatures of the order of 1400°C were optimal. The use of lower temperatures led to the formation of polycrystalline aggregates, while higher temperatures increased the reaction rate so much that silicon carbide was formed in the supply tube.

TABLE 1. Principal Properties of Silicon Carbide

SiC modification	Hardness, kg/mm²	Hardness (mineralogical scale)	Density, g/cm³	Thermal expansion coefficient, deg⁻¹	Thermal conductivity	Lattice parameters, Å	Forbidden band width, eV	Impurity ionization energy, eV — Group V donors	Impurity ionization energy, eV — Group III acceptors	Mobilities, cm²·V⁻¹·sec⁻¹ — Electrons	Mobilities, cm²·V⁻¹·sec⁻¹ — Holes	Effective masses, m_0 — Electrons	Effective masses, m_0 — Holes
α	3500	9.5–9.75	—	$(4-7) \cdot 10^{-6}$	—	$a = 3.08065$ $c = 15.11739$	2.86 at 300°K	0.08	0.27	100	100	0.6	1.2
β	—	—	—	—	—	$a = 4.349$	2.2	—	—	—	—	—	—

Needle-shape crystals of β-SiC were obtained; they were up to 6 mm long and yellow, green, gray, or black. Individual crystals were colorless.

The use of other initial reagents in systems having other construction also yielded small (fraction of a millimeter, or at best several millimeters) crystals of β-SiC in the form of needles, rods, or polyhedra.

It is not yet possible to conclude which of the initial mixtures is the best. The further development of this method requires thorough investigations because many details of the processes of growing silicon carbide crystals from the gaseous phase using chemical reactions are still not clear.

Properties of Silicon Carbide

Physicochemical Properties. The presence of only one compound, SiC, in the Si−C system has been established. The phase diagram of the Si−C system has been given by Nowotny et al. [166]. This diagram has been used to plot the diagram given in Fig. 30, which also includes the regions of the existence of the α- and β-modifications of silicon carbide in accordance with the data given by Wright and Bartels [167]. Both the cubic and the hexagonal forms of SiC are stable [168], but the true position of the boundary between the regions of stable α- and β-SiC is not known. Usually, the cubic modification is formed below 2000°C, while the hexagonal crystals grow at higher temperatures. However, in some cases [164], cubic SiC crystals have been obtained at 2700°C, and hexagonal crystals at 1400-1500°C.

Under normal pressure, silicon carbide melts incongruently, decomposing at 2700°C with the precipitation of graphite.

The problem of the solubility of carbon and silicon in silicon carbide is not clear. Judging by the data from electrical measurements, silicon does dissolve to some extent in silicon carbide. However, the solubility occurs in such very low concentrations that it cannot be detected by chemical means [169].

As mentioned at the beginning of the present section, silicon carbide exhibits polytypism. At present, over 15 polytypic forms of SiC are known. Their formation is associated with the dislocation mechanism of crystal growth.

Various polytypic forms of SiC result from the unequal displacements of silicon (or carbon) atomic layers with respect to one another, these displacements following definite laws. Several layers, which give the whole lattice if repeated along the C axis, form a "stack." Thus the thickness of a stack determines the period of hexagonal silicon carbide. More detailed information on the polytypic forms of silicon carbide is given in [91].

The lattice period of cubic β-SiC is $a_0 = 4.349$ Å, while in the polytypic forms of α-SiC the lattice period along the C axis is 2.5 n, where n is the number of layers in a stack.

The chemical properties of technical-grade silicon carbide are described in detail in [170]. They are characterized by high chemical

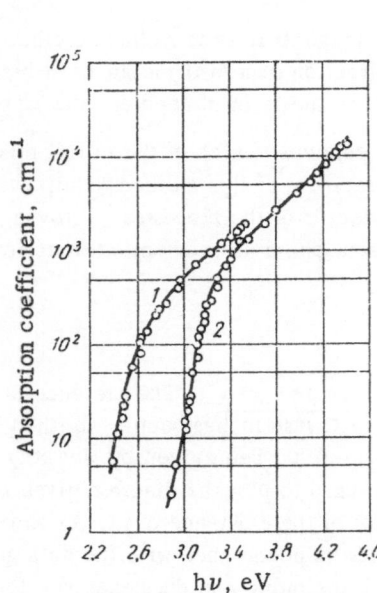

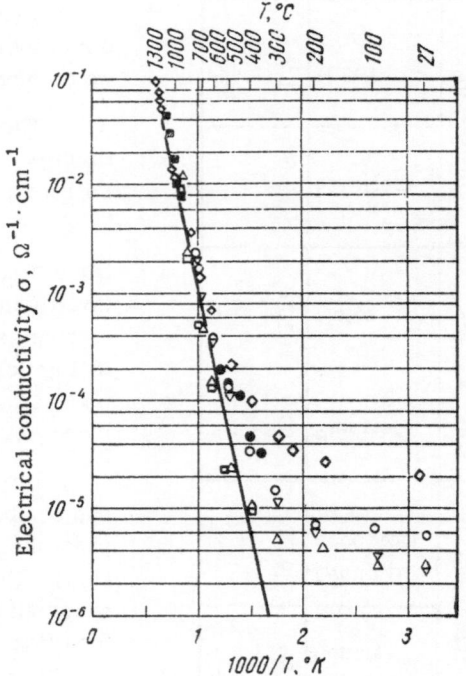

Fig. 33. Absorption spectra of cubic
(1) and hexagonal (2) (6H type)
SiC at 300°K.

Fig. 34. Dependence of the electrical
conductivity of β-SiC on the reciprocal
of temperature.

stability and insolubility in acids with the exception of mixtures of nitric and hydrofluoric acids, phosphoric acids, and alkali melts. Hydrogen, nitrogen, and carbon monoxide do not react with SiC even at very high temperatures. Chlorine begins to react at 400-600°C and it decomposes SiC completely at 1200°C, giving rise to $SiCl_4$ and CCl_4. When heated in air, silicon carbide begins to oxidize slowly at 800°C.

Data on the density, hardness, and other properties of silicon carbide are listed in Table 1.

The hardness of silicon carbide at 25°C is 3500 kg/mm² (9.5-9.75 on the mineralogical scale) [171]. The lattice parameters of the hexagonal modification at room temperature are: a = 3.08065 Å, c = 15.11739 Å[172]. The ratio c/a increases with temperature from 4.907 at −190°C to 4.908 at 750°C, decreasing again to 4.907 at 1200°C. The thermal expansion coefficient of the cubic modification is zero at a temperature close to −273°C; at +300°C it amounts to $4.5 \cdot 10^{-6}$ deg^{-1} increasing up to $5.4 \cdot 10^{-6}$ deg^{-1} at 1200°C.

Measurements of the thermal conductivity of the cubic and hexagonal forms of silicon carbide in the range 10-200°K [173] have shown that the conductivity (in units of $W \cdot cm^{-1} \cdot deg^{-1}$) depends on temperature in the following way.

For the α-modification $\chi_\alpha = 1.1 \cdot 10^{-3} T^{2.0}$ at T < 30°K; $\chi_\alpha = 23 \cdot T^{-0.65}$ at 110 < T < 200°K.

For aggregates consisting mainly of crystals of the β-modification, $\chi_\beta = 1.4 \cdot 10^{-4} T^{2.7}$ at T < 30°K.

Silicon Carbide as a Semiconductor. Silicon carbide belongs to that group of refractory semiconductors whose properties have been investigated in greater detail than those of other refractory semiconducting materials, although in some cases the data are to a certain extent contradictory, which is due to the technical difficulties arising in the preparation of pure silicon carbide with controlled amounts of known impurities.

We shall consider next the band structure of silicon carbide. Several authors [174] agree that there is a maximum of the valence band of SiC at the point $\vec{k} = 0$, as in the band structures of germanium, silicon, and diamond. If, by analogy with these other semiconductors, we assume that the structure of the conduction band of SiC is similar to that of the conduction bands of germanium and silicon, then the absolute minima in the

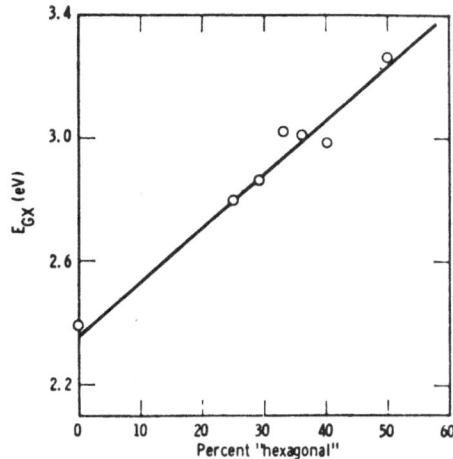

Fig. 35. Exciton width of the forbidden band of various SiC polytypes at 4.2°K as a function of the hexagonal fraction in the lattice.

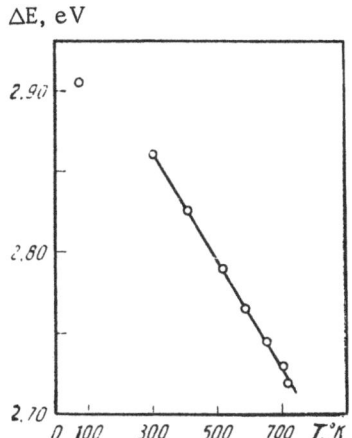

Fig. 36. Temperature dependence of the forbidden band width of α-SiC.

conduction band should be shifted away from the point $\vec{k} = 0$. These considerations can be checked experimentally by investigating the fundamental absorption edge. Measurements of the fundamental absorption edge of cubic silicon carbide (β-SiC) [175, 176] gave a forbidden band width ~2.2 eV (Fig. 33), which was in good agreement with the band structure calculations [177].

Recent measurements [323] of the absorption coefficient and luminescence of single-crystal samples of cubic silicon carbide and of the various polytypes of the hexagonal modification, carried out at about 6°K, established that the transitions in SiC are indirect and involve the participation of four phonons, whose energies are given in the table below.

Phonon mode	Phonon energy (in meV) in polytype		
	Cubic	15R	6H
Transverse acoustic (TA)	46.3	46.3	46.3
Longitudinal acoustic (LA)	79.4	78.2	77.0
Transverse optical (TO)	94.4	94.6	94.7
Longitudinal optical (LO)	102.8	103.7	104.2

It is evident from the above table that the phonon energy depends weakly on the type of the silicon lattice, which allows us to conclude that the energy-band structure symmetry is identical in different polytypes of SiC.

Bassani and Yoshimine [324] calculated the band structure of SiC and showed that a conduction band minimum occurs at the point X of the Brillouin zone, i.e., at the edge of this zone along the < 100 > directions. This theoretical model does not contradict the experimental data.

The same authors determined the "exciton" width of the forbidden band* for various polytypes of SiC at 4.2°K; the results of the measurements are given in the table below.

Polytype	Cubic	8H	21R	15R	33R	6H	4H
Exciton forbidden band width	2.390	2.90	2.86	2.986	3.01	3.023	3.263
"Hexagonal" fraction	0	25	29	40	36	33	50

It is interesting that the forbidden band width is a linear function of the percentage fraction of the "hexagonality" in the lattice of a given polytype; this can be seen in Fig. 35 (for details cf. [325]).

Measurements of the forbidden band width of the cubic modification of silicon carbide, based on the temperatude dependence of the electrical conductivity, give a somewhat different result, namely, 1.90 ± 0.10 eV (Fig. 34).

Hexagonal silicon carbide (α-SiC) has been investigated quite thoroughly. The forbidden band width has been measured very accurately by optical and electrical methods. An investigation of the absorption of light in α-SiC (6H) in the photon energy range from 2.6 to 3.3 eV at temperatures of 77-717°K [178] has shown that the forbidden band width of α-SiC at 300°K amounts to 2.86 eV and that the interband transitions are indirect, involving the absorption or emission of 0.09 eV phonons (Fig. 35). These data are in good agreement with the results of other authors, deduced from measurements of the optical absorption [179], as well as from investigations of the radiative recombination [180] and of the forward branches of the current — voltage characteristics of p-n junctions [144]; they are also in good agreement with the electrical measurements [181]. The forbidden band width decreases linearly as the temperature rises at the rate of $\beta = -3.3 \cdot 10^{-4}$ eV/deg in the range 300-700°K (Fig. 36). At lower temperatures, the coefficient β is smaller.

There are as yet very few data on the influence of impurities on the electrical properties of silicon carbide. This is because it is more difficult to control impurities in SiC than in silicon or germanium. Impurities of group V elements act as donors, while those of group III elements act as acceptors. It is also known that the quantitative data from chemical analyses agree with the results of the electrical measurements. This indicates that each impurity atom of group III and V elements has only one level in the forbidden band of silicon carbide. The ionization energies of nitrogen donor centers are ~ 0.08 eV [182] and the aluminum level lies 0.27 eV above the top of the valence band [183]. According to the same authors, silicon carbide can have one more acceptor level (impurity unknown), which is 0.39 eV above the top of the valence band. Investigations of strongly doped n-type silicon carbide (α-SiC) have shown [184] that silicon carbide exhibits all the effects characteristic of group IV^b semiconductors and $A^{III}B^V$ compounds (the formation of an impurity band, negative magnetoresistance, dependence of the ionization energy of donors on their concentration, etc.). The effects associated with the formation of an impurity band in SiC have also been described in other papers [185, 186].

The mobility of electrons in hexagonal silicon carbide (6H modification), having a carrier density of about 10^{17} cm^{-3}, is more than 200 cm$^2 \cdot$ V$^{-1} \cdot$ sec^{-1} at room temperature and increases with cooling [341]. The hole mobility is usually less than 50 cm$^2 \cdot$ V$^{-1} \cdot$ sec^{-1}. The relatively low mobilities are associated with large effective carrier masses and with the polar scattering, although it is possible that the carrier mobility is low due to a strong compensation of impurity centers. The mobility of electrons in β-SiC has been found to be 32 cm$^2 \cdot$ V$^{-1} \cdot$ sec^{-1} at 300°K [187]. Investigations of the temperature dependence of the electron and hole mobilities in the range 90-1000°K [188] have shown that the electron mobility reaches a maximum at 200°K and then decreases with increasing temperature. At 90°K, the electron mobility is very low. The hole mobility has a similar temperature dependence.

---

*The exciton width of the forbidden band is $E_{Gx} = E_G - E_x$, where E_G is the forbidden band width and E_x is the binding energy of the excitons (it is not known for SiC).

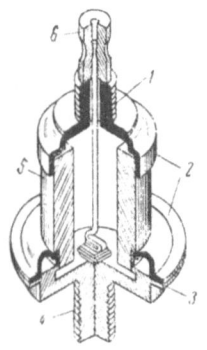

Fig. 37. Section through a silicon carbide rectifier: 1)stainless steel; 2)NiFe; 3)weld; 4) nickel; 5)ceramic; 6)silver.

Estimates of the effective electron and hole masses have yielded the following values [189]: $m_n^* = 0.6\ m_0$, $m_p^* = 1.2\ m_0$.

Investigations of the electroluminescence of SiC are of great interest. The first studies were carried out in 1923 by Losev [190, 191]. Recently, the phenomenon of electroluminescence in SiC has been used to determine the lifetime of nonequilibrium charge carriers [192]. An investigation of the dependence of the radiation intensity when a nonequilibrium carrier pulse is injected into a sample has shown that the lifetime of these carriers does not exceed (even in the purest samples) 0.5 μsec. Recent studies of the photoemf in n- and p-type silicon carbide have revealed that the diffusion length of minority carriers is $\sim 2 \cdot 10^{-3}$ cm [193]. Measurements of the low-frequency permittivity, carried out on cubic and hexagonal SiC using two methods [194, 195], gave similar results. It was found that at low frequencies $\varepsilon = 10.2 \pm 0.2$ while at high frequencies $\varepsilon = 6.7$. The optical dispersion parameters for β-SiC are identical with those for the ordinary ray in α-SiC.

Applications of Silicon Carbide

Silicon carbide has been used for some time in industry, and its applications have included the manufacture of semiconducting devices. The majority of Losev's well-known investigations of the phenomenon of negative resistance in semiconductor detectors (1922-1927) were carried out on silicon carbide. Silicon carbide is used to make nonlinear semiconducting resistors (varistors), used, for example, as dischargers to protect various electrical circuits from overvoltages [196], and for various automatic devices. Silit heaters, thermal compensators, ignitors in ignitrons, and absorptive loads in waveguides are made of silicon carbide [197].

Silicon carbide is now being used to make semiconducting diodes and transistors. The manufacture of such devices requires a suitable raw material (SiC single crystals with known amounts of impurities). Moreover, the processes of the formation of p-n junctions and ohmic contacts (unless, of course, p-n junctions are formed during the growth of the crystals) are very important. The difficulty in providing contacts to SiC is associated with the fact that pressure contacts have high resistance (then the conditions required for the operation of a semiconductor device are absent).

Considerable success has already been achieved in this respect: ohmic contacts with n- or p-type SiC are obtained by fusing-in at 1500-1700°C a suitable silicon alloy (doped with boron or aluminum for contacts with p-type SiC, and with phosphorus or arsenic for n-type) [154, 198, 199] or by fusing-in metal powder mixtures, for example, a mixture of nickel and tungsten powders [154, 199, 200]. Rectifying contacts, i.e., p-n junctions, are obtained by fusing-in p-type silicon into n-type SiC or n-type silicon into p-type SiC [154, 198, 199]. A technique for the chemical treatment of silicon carbide to improve the characteristics of semiconducting devices has been developed in detail [201].

The success in the development of silicon carbide devices is exemplified by p-n junction rectifiers [202]. The crystals were grown in a Lely-type furnace [147], the charge contained an admixture of pure aluminum, and the growth occurred in an atmosphere of pure argon, so that p-type crystals were obtained in the first stage of growth. Then nitrogen was added to the furnace atmosphere and the next part of the crystal was n-type. The central part of the crystal was blue (p-type) and the outer parts were green (n-type). The boundary between these two regions was the p-n junction. After suitable cutting and polishing of the crystal, ohmic tungsten contacts were attached to the n- and p-type regions by fusion [154] and the whole system was placed in a holder. The external view of the resultant rectifier is shown in cross section in Fig. 37. Silicon carbon rectifiers can work at temperatures up to 500-700°C and they can carry currents up to 10A in the forward direction and withstand reverse voltages up to 200 V. Work has also been published [189] on the preparation of silicon carbide transistors and other amplifying elements for radio circuits, capable of operating at frequencies up to 100 Mc at ambient temperatures from −75°C to +1400°C.

It seems quite clear that silicon carbide is one of the most promising semiconducting materials It is essential, then, to improve as quickly as possible methods of preparing large and perfect single crystals, control of the impurity concentration during growth, and technology of the manufacture of materials and devices.

COMPOUNDS BASED ON BORON

The group of semiconducting compounds of the $A^{III}B^V$ type based on boron is one of the least known among the semiconducting materials with the diamond-like structure. However, these compounds are of considerable interest because of their high chemical stability, large forbidden band width, and other special properties. These properties are governed by the special position of boron in the periodic system. Boron belongs to those elements of the second series of the periodic system which have the strongest bonds. While aluminum, gallium, and indium have low melting points, low hardness, and are typical metals, boron is a refractory (melting point at $\sim 2300°C$) and hard ($300 \ kg/mm^2$) semiconductor. From this, we conclude that boron-based $A^{III}B^V$ compounds should have interesting properties.

Cubic Boron Nitride BN (Borazon)

Boron nitride, BN, is the electronic analog of carbon. As a chemical compound, it has been known for over 100 years. Hexagonal boron nitride, which is very similar in structure to graphite (Fig 38), can be obtained in many ways. The graphite-like structure of hexagonal BN suggests that this compound may exist also in another modification with a structure similar to diamond. The existence of a cubic diamond-like form of boron nitride was suggested some time ago [203, 204], but the first report on the preparation of the cubic modification was published as recently as 1957 [205].

The reason for this delay in the preparation of the cubic boron nitride becomes clear if we extend the analogy between carbon and boron nitride to the physicochemical properties of these materials. As mentioned in the section dealing with diamond, the latter is thermodynamically stable only at very high pressures (cf. Fig. 23). In the absence of very high pressures, the stable form of carbon is the hexagonal modification, i.e., graphite. Therefore, one would expect that the stable phase of boron nitride at relatively low pressures is the hexagonal form of BN, and the preparation of the cubic modification of this compound requires the use of high-pressure techniques.

It is, therefore, not surprising that the preparation of cubic boron nitride has only recently become possible, when techniques have been developed so that we can apply pressures of hundreds of thousands of atmospheres at temperatures of several thousand degrees. The need for high temperatures to achieve the allotropic transformation of the hexagonal boron nitride into the cubic form is related, as in the case of the graphite – diamond transformation, to the fact that at relatively low temperatures this transformation is "frozen," i.e., it is so slow that it cannot be realized in practice.

These theoretical considerations have been confirmed by Wentorf [205, 206], whose work is summarized below. Wentorf used a high-temperature technique to prepare boron nitride BN with the zinc blende structure. This cubic form of boron nitride is known as borazon.

Preparation of Borazon

The published methods of preparing cubic boron nitride can be divided into three groups.

The first group includes the methods in which a high-pressure technique is used to produce the allotropic transformation

$$BN \ (hex.) \rightarrow BN \ (cub.)$$

in the presence of catalysts.

The second group embraces the methods in which again a very high pressure is used but the allotropic transformation of boron nitride is replaced with some chemical reaction.

The third group includes the preparation of cubic boron nitride at pressures close to normal.

All these methods will be considered in detail in the present section.

The high pressure needed in the first two groups of methods is established by means of apparatus which is used to prepare synthetic diamonds [116]. A sample, consisting of the initial product and the catalyst, is heated

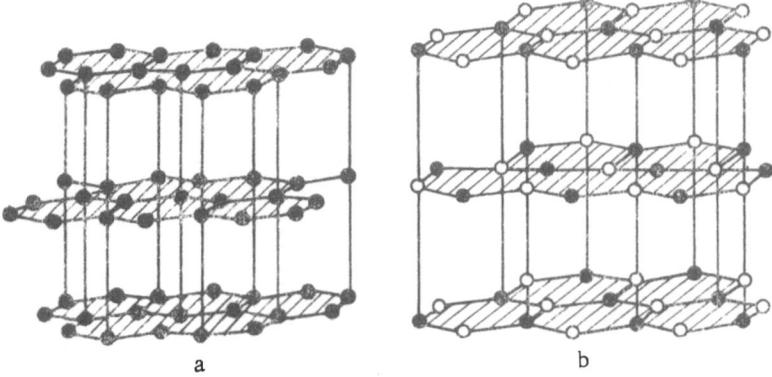

Fig. 38. Structure of graphite (a) and boron nitride (b): ●) B; ○) N.

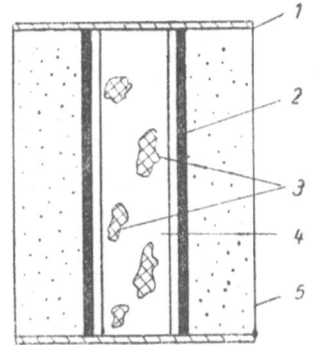

Fig. 39. Schematic representation of a unit which is placed in a high-pressure chamber: 1) tantalum or titanium disk; 2) heating tube (made of graphite, tantalum, titanium, etc.); 3) lumps of "catalyst"; 4) hexagonal BN; 5) pyrophyllite insulator.

by means of a current passing through a heating tube made of graphite, tantalum, etc., placed in a reaction chamber. A typical reaction chamber, placed in a high-pressure enclosure, is shown in Fig. 39. The chamber has a height of 11.5 mm and a diameter of ~9 mm. Using this technique, one can study processes occurring at pressures up to 100,000 atm and temperatures up to 2500°C.

The allotropic transformation BN(hex.)→BN(cub.) requires that the hexagonal boron nitride (containing a catalyst) be kept at a high temperature and pressure for a suitable period. The temperature is gradually reduced until the transformation "freezes" and then the pressure is reduced to atmospheric.

The preparation of the initial material – the hexagonal boron nitride – does not present any special difficulties [207].

In his first attempts to prepare borazon, Wentorf [206] tried to ease the allotropic transformation BN(hex.)→BN(cub.) by using transition metals (iron, nickel, manganese) as catalyst admixtures, i.e., those catalysts which have been found effective in the graphite→diamond transformation. However, the cubic form of BN was not obtained even at 100,000 atm, at temperatures above 2000°C. The sole result was some enlargement (from 5 to 20 mm) of the crystallites of the initial hexagonal boron nitride.

The failure of transition metals as "catalysts" forced Wentorf to search for suitable catalytic materials. He found that alkali and alkaline earth metals, as well as antimony, tin, and lead, were suitable. The use of other elements did not give positive results.

The new catalysts differed in their efficiency and, therefore, depending on the actual catalyst used, the transformation BN(hex.)→BN(cub.) occurred at different pressures (50,000-90,000 atm) and temperatures (1500-2000°C). It was found that the pressure and temperatures needed for the allotropic transformation increased with the atomic weight of the "catalyst" used. In order to achieve the transformation BN(hex.)→BN(cub.) using potassium or barium as a catalyst, a minimum pressure of 70,000 atm is required.

When the pressure was reduced somewhat, borazon was not formed although potassium and barium reacted with the hexagonal boron nitride and diffused into it. On the other hand, when lighter metals – magnesium, calcium, or lithium – were used as catalysts, the formation of the cubic boron nitride was observed at pressures of 45,000 atm and the process was characterized by a high yield of the required product.

It was found also the that efficiency of the catalyst employed dropped markedly in the presence of several percent of water, boron anhydride, and other impurities.

These investigations allow us to conclude that boron nitride, like carbon (Fig. 23), may be stable in two forms: hexagonal and cubic. The region of the stable existence of borazon lies at high pressures and is separated from the region of the existence of the hexagonal boron nitride by a boundary line representing the conditions of the coexistence in equilibrium of the two crystalline modifications of boron nitride. This boundary line, as in the case of carbon, does not run parallel to the abscissa axis (temperature axis) but makes an angle with it so that as the temperature is increased higher pressures are needed in order to achieve the BN (hex.) → BN (cub.) transformation.

A comparison of the boundary line with the corresponding line in the carbon system (Fig. 23) shows that at a given temperature the transformation BN (hex.)→BN (cub.) is observed at a lower pressure than the graphite→diamond transformation.

It was found that nitrides of the metals referred to above could also act as catalysts. Since the use of light metals had certain advantages, lithium, magnesium, or calcium nitrides were used as catalysts. By means of these nitrides it was possible to grow cubic boron nitride crystals at pressures of 44,000-74,000 atm and temperatures of 1200-2000°C. The total amount of borazon formed in one run reached 0.3 g, and the dimensions of individual polyhedral crystals were as much as 0.7 mm [205, 206].

The system boron nitride—lithium nitride has been investigated most. It was established that in this system the formation of a complex with the approximate composition $Li_3N \cdot 3BN$ is possible. "This complex acted as a molten solvent which dissolved the hexagonal BN and forced the precipitation of the cubic boron nitride due to a departure from thermodynamic equilibrium into the region where the cubic form was stable at the working pressure and temperature" [206].

Obviously, this process represented the recrystallization of boron nitride from the melt. $Li_3N \cdot 3BN$ acted as the liquid solvent and boron nitride crystallized from the melt when the melt was supersaturated. The high pressures and temperatures ensured the diamond-like structure of BN crystals precipitated from the melt.

The dimensions of the borazon crystals formed in this way depended primarily on the pressures and temperatures at which the crystals were formed. The closer the pressure and temperature were to the boundary line between the cubic and hexagonal forms of boron nitride, the larger were the resultant crystals. If the process took place at the boundary line but at high temperatures, the rate of reaction increased; even a small change in the pressure or temperature had a strong effect on the quality of the crystals. Therefore, the best crystals were grown at moderate pressures and temperatures (50,000 atm and 1700°C). The time needed to form crystals under these conditions was only several minutes. Upon increasing the pressure to 70,000 atm, the crystals shrank to 0.02 mm, which indicated a considerable increase in the rate of formation of three-dimensional nuclei.

The second group of methods includes chemical reactions which take place at very high pressures [206]. Mixtures of boron and lithium nitride (or magnesium nitride) were used as the initial materials. The formation of the cubic boron nitride was observed but the yield of the process and the quality of the crystals were poorer than in the case of the formation of borazon from the hexagonal boron nitride using a nitride catalyst.

R. C. Vickery [208] described a method of preparing the cubic form of boron nitride at normal pressure by nitriding boron phosphide BP at 800°C. The boron phosphide, obtained by Vickery using the thermal dissociation of coordination compounds of the $BCl_3 \cdot PCl_5$ type, was in the form of thin dark films. When they were treated in a flowing mixture consisting of 5% ammonia and 95% nitrogen, the films became lighter in color and phosphine was liberated. X-ray diffraction analysis of the product obtained revealed that it was the cubic form of boron nitride. Vickery assumed that the process involved the reaction

$$BP + NH_3 \rightarrow BN \text{ (cub.)} + PH_3.$$

In view of the data of Wentorf [206] and the crystallochemical similarity of the cubic form of boron nitride and of diamond, Vickery's claims must be regarded with caution, especially as there has not been a single confirmation of his method of preparing borazon.

TABLE 2. Principal Properties of $A^{III}B^{V}$-type Compounds Based on Boron (sphalerite structure)

Compound	T_{mp}, °C	Heat of formation, kcal/mole	Lattice period, Å	Density, g/cm³	Micro-hardness, kg/mm²	Hardness on Mohs' scale	Forbidden band width, eV	Thermo-electric power, μV/deg
BN (cub.)	~3000	—	3.615	3.45	—	10	~5 (theoret.)	—
BP	~2500	49	4.538	2.89	3700	—	5.9	300
BAs	~2000	—	4.777	—	~1900	—	~3 (theoret.)	49

Properties of Borazon

Physicochemical Properties. The phase diagram of boron nitride, like that of the B−N system, is not yet known. There are indications that, apart from boron nitride BN, the system includes other compounds of boron and nitrogen: boron triazide $B(N_3)_3$ [209] and, possibly, a lower boron nitride B_3N [210].

A chemical analysis of borazon crystals showed that they contained 41.5 wt.% boron and 50.1 wt.% nitrogen (the theoretical composition of BN is: 43.6% boron and 56.4% nitrogen). This analysis was carried out by dissolving borazon in molten NaOH (the ammonia formed by this reaction was titrated), since none of the usual acids reacts with borazon [205].

Borazon crystals are not affected when heated in a vacuum to temperatures above 2000°C. When heated in air, borazon begins to oxidize slowly only at 2000°C, while diamond burns in air at 875°C.

Borazon heated under 40,000 atm transforms into the hexagonal modification of boron nitride at 2500°C. Borazon crystallizes the zinc blende (sphalerite) structure with a lattice period 3.615 ± 0.001 Å at 25°C.

Borazon crystals, obtained after the allotropic transition BN (hex.)→BN (cub.) at high pressures, are in the form of polyhedra, usually tetrahedra and octahedra. They are transparent and their color depends on the impurities present. Thus, boron colors borazon crystals brown or black, beryllium makes them blue, and sulfur yellows them. Yellow borazon crystals are also obtained from a mixture of hexagonal boron nitride with lithium nitride. Red, white, and colorless crystals can also be obtained.

The density of borazon is 3.45 g/cm³ (the x-ray density is 3.47 g/cm³). The hardness of borazon on Mohs' scale is comparable with that of diamond (10).

The data on some properties of boron nitride are presented in Table 2.

Electrical and Optical Properties. Measurements of the reflection spectrum of cubic boron nitride (obtained using lithium nitride as a "catalyst") have shown [206] that these crystals absorb strongly in the 7-9 eV range. The general form of the absorption spectrum is similar to the absorption spectrum of diamond (cf. Fig. 25) but is characterized by energies about 2 eV greater than those for diamond. The refractive index of cubic boron is 2.22.

The band structure of cubic boron nitride has been calculated theoretically [76] from the band structure of diamond by the perturbation theory methods (Fig. 40). The maximum of the valence band remains at the point $\vec{k} = 0$ but the minimum of the conduction band is displaced along the [100] axis. Consequently, the value of ΔE increases to ~10 eV, which is double the value of ΔE for diamond.

An investigation of the influence of impurities on the conductivity of borazon showed that the presence of metallic beryllium or its salt (from 0.01 to 1 wt.%) in the reaction mixture tended to yield BN crystals having p-type conduction. Such crystals had a resistivity of 10^3 Ω·cm, although sometimes resistivities as low as $2 \cdot 10^2$ Ω·cm were observed at room temperature. The value of the activation energy for conduction varied from crystal to crystal (it could have been affected by contacts). From experiments on the doping of

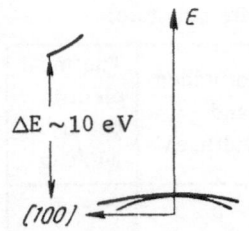

Fig. 40. Band structure of the cubic form of boron nitride.

other $A^{III}B^V$-type compounds, Wentorf [206] concluded that beryllium atoms may replace boron or nitrogen atoms in the lattice of the cubic modification of boron nitride.

Attempts to establish p-type conduction in borazon crystals, obtained from the B—N—Li system, by adding magnesium or zinc to the reaction mixture were not successful. This was probably due to the relatively large dimensions of the admixture atoms and the consequent difficulty of replacing the atoms in the borazon lattice.

N-type conduction was obtained in borazon crystals by adding to the reaction mixture an excess of boron, sulfur, silicon, KCN, etc. An admixture of boron made the resultant crystals dark brown. Crystals prepared in this way had a high resistivity.

The most active donor was sulfur. With the addition of 0.3-3% sulfur to the initial mixture, it was possible to obtain crystals whose resistivity was 10^4 $\Omega \cdot$ cm and sometimes even 10^5 $\Omega \cdot$ cm at 25°C. Wentorf suggested that sulfur atoms replaced nitrogen atoms in the cubic lattice of boron nitride. The ionization energy of impurity centers was 0.05 eV.

When compounds containing carbon and nitrogen were added to the reaction mixture, Wentorf was able to obtain crystals having n-type conduction, 10^5-10^7 $\Omega \cdot$ cm resistivity, and a conduction activation energy of 0.28-0.41 eV. These crystals were yellow, brown, or reddish brown.

N-type conduction was sometimes observed for borazon crystals obtained from the reaction mixtures lithium nitride—boron nitride or magnesium nitride—boron nitride without any deliberate addition of admixtures. These crystals usually had high resistivity — of the order 10^6-10^9 $\Omega \cdot$ cm at room temperature. The cause of the n-type conduction in this case was oxygen, which was very difficult to eliminate from the reaction mixtures due to the high activity of the nitrides present in the charge. This was in agreement with the observations that, when magnesium nitride, which was a stronger reducing agent than lithium nitride, was used as a catalyst, boron crystals had a higher resistivity.

An investigation of the rectifying properties of crystals of the cubic modifcation of boron nitride was carried out on a pair of n- and p-type crystals which were in contact. A weak constant current (10^{-6} A) was passed through this pair under a low voltage (5 V), using silver contacts. The ratio of the forward to reverse current was quite low: from 2 to 20. At 25°C, the strongest currents were passed so that the p-type crystal was positive. At temperatures of 300-400°C, the direction of rectification was sometimes reversed in some crystal pairs. On cooling back to room temperature, the initial direction of rectification was re-established.

Further progress in the investigation of the properties of the cubic modification of boron nitride requires the preparation of larger crystals, of suitable shape, and improvements in the technology of p-n junction preparation.

Applications of Borazon

In view of the wide forbidden band and the relatively low ionization energy of impurity centers (0.05 eV, as in silicon), we would venture to suggest that in the future borazon will be found to be an extremely convenient material for semiconducting devices working in a wide range of temperatures. Unfortunately, at present, we know too little about its physical properties; we do not know the electron and hole mobilities or the effective carrier masses. Therefore, before drawing any final conclusions about the applications of borazon in the manufacture of semiconducting devices (which may be economic in spite of the technological difficulties of preparing the cubic modification of BN), this semiconductor must be investigated in much greater detail.

Boron Phosphide BP

Many semiconducting binary compounds were synthesized long before the science of semiconductors became established. This happened to the majority of $A^{III}B^V$ compounds based on aluminum, gallium, and indium. For example, many compounds of this type, such as aluminum phosphide, gallium arsenide, indium antimonide, and others, were known in the 1920's.

The development of semiconductor science and the practical applications of substances exhibiting semiconducting properties have given a new lease of life to these substances. This applies, in particular, to boron phosphide BP, which was first prepared by the French chemists, Besson and Moissan (independently of each other), in 1891. Between them they published three papers in which they described the methods of preparation and some chemical properties of this interesting compound. Sixty years later, new interest was aroused in boron phosphide, this time in connection with its semiconducting properties. Boron phosphide, like other $A^{III}B^V$ semiconducting compounds, is formed from elements of groups III and V of the periodic system.

The first papers dealing with boron phosphide BP as a semiconducting compound appeared in 1957. Since then, upward of ten papers have been published about this compound. This number is very small compared with the many hundreds of papers dealing with other semiconducting compounds. Evidently, the difficulties which one meets in the preparation of this compound in the form of a polycrystalline ingot or as a single crystal have held up progress. Nevertheless, the interest in boron phosphide BP has increased every year and there is no doubt that methods of preparing crystals of this compound will be developed soon.

Preparation of Boron Phosphide

A large number of methods of preparing boron phosphide BP have been described. These methods may be divided into direct and indirect methods of synthesis.

Direct Synthesis from Elements. This is one of the simplest and most widely used methods of preparing boron phosphide BP. It was first reported in 1957 [211, 212]. Later investigations have also used this method [213, 214].

Boron phosphide BP is formed by the reaction

$$2B + P_2 \rightarrow 2BP.$$

The reaction is exothermic, but is characterized by a considerable activation energy, so that quite high temperatures are needed to make it proceed in the forward direction.

There are two possible ways of realizing direct synthesis: in a two-temperature furnace or in a single-temperature furnace. In both cases, the initial substances are usually amorphous boron and red phosphorus. Use of the so-called compacted boron is also possible but one would expect that phosphorus would react with this type of boron much more slowly. Moreover, when the reaction is not complete, one may obtain samples of intermediate composition, $B_xP_{(1-x)}$, where $x > 0.5$. Such samples contain an excess of boron, which is difficult to remove.

The simplest way of achieving direct synthesis is in a two-temperature furnace. The charge of boron and phosphorus (taken in the stoichiometric ratio or with a small excess of phosphorus) is placed at the hot end of an evacuated and sealed quartz ampoule. The hot zone of the furnace may be raised rapidly to 1000-1100°C, and the cold zone to 420-450°C. This establishes a phosphorus vapor pressure of 1-2 atm in the ampoule. Under these conditions, the reaction of boron with phosphorus is quite slow and in order to increase the yield of boron phosphide BP to 85% (at a phosphorus vapor pressure of the order of 1-2 atm), about 50 hours are needed [214].

In order to increase the rate of reaction, it must be carried out at higher temperatures, but this involves the danger that boron will reduce the silicon present in the quartz.

The direct synthesis may be carried out also in a single-temperature furnace. However, the heating cannot be undertaken quickly because there is a danger of explosion due to the high vapor pressure of phosphorus. The temperature should be kept low so that the vapor pressure of the phosphorus in the ampoule is less than the mechanical stress of the latter. On the other hand, high temperatures are needed to increase the reaction rate in the ampoule. These two requirements determine some optimum rate of heating in the furnace. The temperature range in which P_4 molecules dissociate into diatomic molecules (this process begins from about

800°C) is also dangerous. Such dissociation also increases the vapor pressure of the phosphorus in the ampoule and may explode the latter. This has been observed at 800-850°C. The product of the direct synthesis is a brown lump, whose individual grains are lightly sintered. Under a light pressure, such a lump breaks up into a powder. The color of the product depends on the amount of free boron. If this amount is large, the product is dark brown; as the boron content is decreased, the color becomes lighter. Relatively pure boron phosphide is light brown. It is not difficult to separate boron phosphide from the excess phosphorus and boron. The phosphorus may be removed directly after synthesis by heating to 400-500°C that part of the ampoule which contains the product. This drives off the excess phosphorus (or phosphorus which has not reacted) to the cold end of the ampoule.

Boron phosphide may be separated from unreacted boron by treating the reaction product in nitric acid, which dissolves boron but does not affect boron phosphide. Boron phosphide may be obtained not only in the form of a powder but also in compacted form. For this purpose, it is necessary to use compacted boron rather than the amorphous element.

Another method of obtaining compacted boron phosphide is to use vibration mixing. In this case, a single-temperature furnace is employed and a stoichiometric charge of amorphous boron and red phosphorus is used. The application of 100 cps vibrations of 0.2-0.3 mm amplitude to an ampoule during synthesis [215] gives rise to the formation of compacted boron phosphide, so hard that it can scratch the quartz. However, the dimensions of the crystals in the compacted boron phosphide (as in the loose-grained product) are very small: less than 1 μ.

Thermal Dissociation of Various Compounds. In this type of method, boron phosphide is prepared using various compounds which dissociate at high temperatures and which contain boron and phosphorus. The common features of these methods are relatively low temperatures needed to prepare boron phosphide, and the amorphous state of the resultant product.

These methods were used by the discoverers of boron phosphide, Besson and Moissan. Besson [216], who also investigated the properties of coordination compounds, discovered that when $BBr_3 \cdot PH_3$ was heated to 300°C, it decomposed, producing hydrogen bromide. The solid brown residue was BP. Thus, the reaction of the dissociation of $BBr_3 \cdot PH_3$ may be represented by the equation

$$BBr_3 \cdot PH_3 \rightarrow BP + 3HBr.$$

Boron phosphide may also be obtained in the same way from the compound $BCl_3 \cdot PH_3$.

A similar process was described in 1958 [208]. The initial compound was $BCl_3 \cdot PCl_5$. It was placed in an ampoule which was evacuated and sealed. At 300°C, thermal dissociation occurred producing chlorine and throwing a deposit onto the walls of the ampoule. When the ampoule was broken, the chlorine escaped and the undissociated residue of $BCl_3 \, PCl_5$ was removed by washing out with water. This gave rise to a thin film, floating on water, which (according to x-ray and chemical analyses), was boron phosphide BP. The processes of the thermal dissociation of $BBr_3 \cdot PH_3$ and $BCl_3 \cdot PH_3$ were investigated by the same author.

The dissociation of boron iodophosphide BPI_2 is described in [217]. If BPI_2 is heated in a stream of hydrogen, then at 160°C one iodine atom is lost and the compound becomes BPI. On further heating to 450-500°C, BPI loses another iodine atom and becomes BP.

As mentioned earlier, all these methods yield boron phosphide BP in the amorphous state. X-ray diffraction patterns of this substance have no sharp reflections, characteristic of crystalline BP, and its chemical stability is relatively low.

The use of these methods to prepare boron phosphide BP can hardly be recommended because they require, as a preliminary step, the synthesis of coordination compounds or boron iodophosphide. If crystalline boron phosphide BP is formed, it is relatively easy to separate it from other reaction products or from unreacted initial components; this is because crystalline boron phosphide has very high chemical stability. However, if amorphous BP is formed, as in the reaction described above, it may be difficult to separate it from the resultant mixture.

The processes described here can be used to prepare thin amorphous films of boron phosphide BP [217].

Other Indirect Methods.   This group of methods of preparing BP includes a great variety of chemical reactions.

We shall consider a method based on the substitution reaction with zinc phosphide, Zn_3P_2 [218]:

$$Zn_3P_2 + 2B \rightarrow 2BP + 3Zn.$$

The reaction is carried out in an evacuated and sealed quartz ampoule. The ampoule is placed in a single-temperature furnace, which may be heated to 900-1100°C. The rate of heating may be quite high because the vapor pressure of Zn_3P_2 at these temperatures is not too high. At 900-1100°C, about 5-10 hours are required. The yield of boron phosphide under these conditions is 80-95%. The next operation is the vaporization of the zinc which is formed by this reaction, and of the unreacted zinc phosphide. Since zinc boils at 907°C, and the rate of sublimation of Zn_3P_2 becomes considerable at 1000-1100°C, it is convenient to carry out this stage by heating the reaction product to the latter temperature. In practice, the ampoule in which the synthesis of BP has been carried out is placed in the furnace in such a way that the end which does not contain boron phosphide is in air. Then the furnace is heated to the required temperature and this drives the zinc and Zn_3P_2 to the cold part of the ampoule where they solidify. Brown boron phosphide powder, mixed with unreacted boron, remains at the hot end of the ampoule. The separation of BP from boron may be carried out in exactly the same way as in the case of direct synthesis from elements.

Boron phosphide prepared by this method is a fine-grained powder, which does not differ greatly from boron phosphide formed by direct synthesis from its elements.

Boron phosphide may also be obtained by the reaction of BCl_3 or BBr_3 (in an argon stream) with Zn_3P_2 at 1000°C [335].

Williams and Ruehrwein [214] have described the preparation of boron phosphide BP by the reaction of boron trichloride and phosphine in accordance with the equation

$$BCl_3 + PH_3 \rightarrow BP + 3HCl.$$

Phosphine was mixed with boron trichloride and the mixture was passed through a quartz tube heated to 1000°C. The reaction produced a lustrous black film on the tube walls; analysis showed that this film was mainly boron phosphide BP. Since phosphine is a poisonous gas, which dissociates at high temperatures, Williams and Ruehrwein used a mixture of BCl_3 with an excess of hydrogen, which was passed over red phosphorus heated to a temperature that ensured the required phosphorus vapor pressure. The resultant mixture of BCl_3, hydrogen, and phosphorus was passed through a quartz tube heated to 1000°C, where boron phosphide was formed in accordance with the reaction

$$2BCl_3 + P_2 + 3H_2 \rightarrow 2BP + 6HCl.$$

This process should not differ basically from the preceding reaction because phosphine PH_3 dissociates into its elements at temperatures above 400-500°C. Williams and Ruehrwein mention that this reaction has several stages with a probable intermediate product of the $BPCl_2$ type. This is confirmed by the fact that when phosphorus was excluded from the reaction mixture, boron phosphide was not formed. On the other hand, it was not possible to prepare BP from BCl_3 and phosphorus alone.

Williams and Ruehrwein also described a method for preparing boron phosphide by the reaction

$$BCl_3 + AlP \rightarrow BP + AlCl_3.$$

Boron trichloride was passed over aluminum phosphide, heated to 975°C. The reaction product was boron phosphide mixed with some aluminum oxide Al_2O_3 (about 5-10%), contained in the original AlP.

It is very difficult to prepare boron phosphide crystals from a stoichiometric melt. This is because of the refractory nature of boron phosphide and its dissociation at high temperatures.

The melting point of boron phosphide has not yet been established. However, it can be estimated approximately from relationships which govern the properties of its analogs. Using this approach, it has been estimated that the melting point of BP should be between 2000 and 3000°C [79].

There are no exact data on the dissociation pressure of boron phosphide. Williams and Ruehrwein [214] give a temperature dependence of the dissociation pressure of BP but they point out that their data may be in error by one order of magnitude:

$$P_P = \frac{13.7 \cdot 10^3}{T} + 10.1 \text{ mm Hg}.$$

At 1625°C, the vapor pressure of phosphorus is 1 atm (however, bearing in mind the possibility of error by one order of magnitude, the vapor pressure of phosphorus above boron phosphide at 1625°C may range from 0.1 to 10 atm). Thus, at the melting point of boron phosphide, its dissociation pressure may reach many hundreds of atmospheres. It is thus clear that the preparation of a stoichiometric melt of boron phosphide requires temperatures higher than 2000°C, and the phosphorus vapor pressures of several hundred atmospheres present an extremely difficult problem. Therefore, to prepare boron phosphide crystals, one should turn to the other methods: growing crystals from solutions or from the gaseous phase.

The technique of crystallization from solutions has been employed by several workers [219, 220] to prepare small boron phosphide crystals using the systems B−P−Ni and B−P−Fe.

Stone [220], who used these systems, took out a patent for a method of growing BP crystals, based on the saturation with phosphorus of a melt consisting of 93 wt.% nickel and 7 wt.% boron. The boron−nickel mixture was placed in a graphite boat, which was located in a mullite tube. This tube was evacuated to 10^{-5} mm Hg and then white phosphorus was distilled in it from a special reservoir. The boat temperature was gradually lowered (at a rate of 5 deg/hr) to the temperature of solidification of the liquid (1170°C). After cooling to room temperature, the final product was dissolved in nitric acid, which produced red BP crystals of up to 2-3 mm in size.

The process was similar when the B−P−Fe system was used.

The crystallization technique described by Stambaugh, Miller, and Himes [221], using a temperature gradient across a boat, may be more promising. Here, the crystallization is directional; the increase in temperature of the crystallization front, moving toward higher temperatures, is compensated by an increase in the phosphorus vapor pressure during the process (by increasing the temperature of the evaporator which is the source of phosphorus). Thus, the required degree of supersaturation of the solution at the crystallization front is maintained during the process.

There are other ways of growing boron phosphide crystals from molten solutions, which are similar to the "classical" methods, used to grow crystals from stoichiometric melts, as well as methods involving various ways of feeding the melt with a solid or with the vapor of a volatile component.

Boron phosphide crystals can also be grown from the gaseous phase. A brief communication has been published on the possibility of growing small needle-shaped or plate-like BP crystals (measuring 1-2 mm) by a gaseous-phase process using a flow system [222]. In this technique, hydrogen chloride or hydrogen bromide is passed over amorphous or microcrystalline boron phosphide at a rate varying from 1 to 1000 milliliter/min. Boron phosphide, heated to 800-1200°C, reacts with the HCl (or HBr), producing a certain amount of volatile products which is carried to the hot zone (1100-1500°C), where the usual reaction takes place, giving rise to boron phosphide, growing in the form of needle-shaped transparent red crystals.

In conclusion, we may say that the gas-phase methods and the methods using the technique of molten solutions are very promising. Although investigations of their use in growing boron phosphide crystals have just begun, there is no doubt that large BP crystals will soon be produced using one of these techniques.

Properties of Boron Phosphide

Physicochemical Properties. The phase diagram of the B−P system is not known yet in full. Apart from boron phosphide, it is known that a second phosphide of composition $B_{13}P_2$ is formed.

In considering the chemical properties of boron phosphide, it is necessary to point out that the properties of the amorphous boron phosphide differ considerably from those of the crystalline modification. We shall deal first with the properties of the amorphous BP, which are described in detail in [216, 217].

Externally, amorphous boron phosphide resembles a brown powder. It is not sublimated in vacuum by heating to 500°C but burns in oxygen at 200°C. At 400°C, amorphous boron phosphide is decomposed by water vapor, and at red heat is decomposed by hydrogen sulfide. It burns at room temperature in chlorine, forming BCl_3 and PCl_5, and interacts rapidly with bromine and iodine vapors. Sulfur vapor converts BP into boron sulfide and phosphorus sulfide. Hydrogen fluoride and hydrogen chloride react with BP under the influence of heat. Amorphous BP burns in ammonia at 300°C, forming boron nitride and phosphorus.

Amorphous BP does not dissolve in water. It burns in nitric acid monohydrate and dissolves in nitric and sulfuric acids when heated. Hydrochloric and hydriodic acids do not react with BP. It decomposes slowly in concentrated solutions of alkalis when heat is applied, and rapidly in alkali melts. The amorphous boron phosphide burns in molten alkali nitrates.

Heating the amorphous boron phosphide in a stream of hydrogen or nitrogen decomposes it and partly liberates phosphorus. The residue still contains boron and phosphorus even if the heating is continued for many hours. The residual compound is found to be B_5P_3, but later investigations have failed to confirm the existence of such a compound of boron and phosphorus.

Crystalline boron phosphide BP has completely different chemical properties.

Microcrystalline boron phosphide, like the amorphous modification, is a brown powder, while large BP crystals are transparent and are reddish brown or orange-red.

The majority of investigators who have prepared the crystalline phosphide have remarked on its unusual chemical inertness. It is not dissolved by any of the known mineral acids or their mixtures, not even after prolonged boiling. It has been found impossible to dissolve boron phosphide in concentrated aqueous solutions of alkalies or in various semiconductor etchants. Only molten NaOH reacts slowly with crystalline BP.

When crystalline boron phosphide is heated it resists oxidation right up to 800-1000°C [214].

On heating crystalline boron phosphide in vacuum or in an inert atmosphere at temperatures exceeding 1100°C, it is found that phosphorus is liberated and the phosphide BP is transformed into a boron phosphide of different composition. There is a considerable divergence of opinions on the composition of this new phosphide. Moissan has ascribed the formula B_5P_3 to the new phosphide [217]. Williams and Ruehrwein [214] are of the opinion that the product of the dissociation of BP is hexaboron phosphide B_6P. Matkovich [223], who investigated in detail the dissociation product, ascribes to it the formula $B_{13}P_2$.

There are no published thermodynamic data on boron phosphide. There is only a theoretical calculation of its heat of formation, H = 49 kcal/mole [224].

Boron phosphide BP crystallizes out with the zinc blende (sphalerite) structure. Rundquist [219] points out that BP represents the structure type B3 and belongs to the space group $F\overline{4}3m$. The value of the lattice constant has been determined by many investigators: 4.55 [211]; 4.537 [212, 214]; 4.538 [213, 219].

The density is 2.89 g/cm^3. This value is slightly less than the density found from the x-ray data: 2.97 g/cm^3 [212].

Boron phosphide is exceptionally hard. There are no reliable data on its hardness, but it is known that it scratches tungsten carbide [211], quartz, agate, and sapphire [214]. Stone and Hill [72] have reported that the hardness of boron phosphide is approximately equal to the hardness of silicon carbide SiC (the hardness of SiC measured with Knoop's pyramid is $H_{100} = 3200$ kg/mm^2). Valov and Lubenskaya [218] found the hardness to be 3700 kg/mm^2.

Electrical and Optical Properties. There have been few investigations of the properties of boron phosphide. The main source of information on some of the electrical and optical properties is the paper of Stone and Hill [72]. The measurements were carried out on BP crystals (measuring 1-2 mm) and on films.

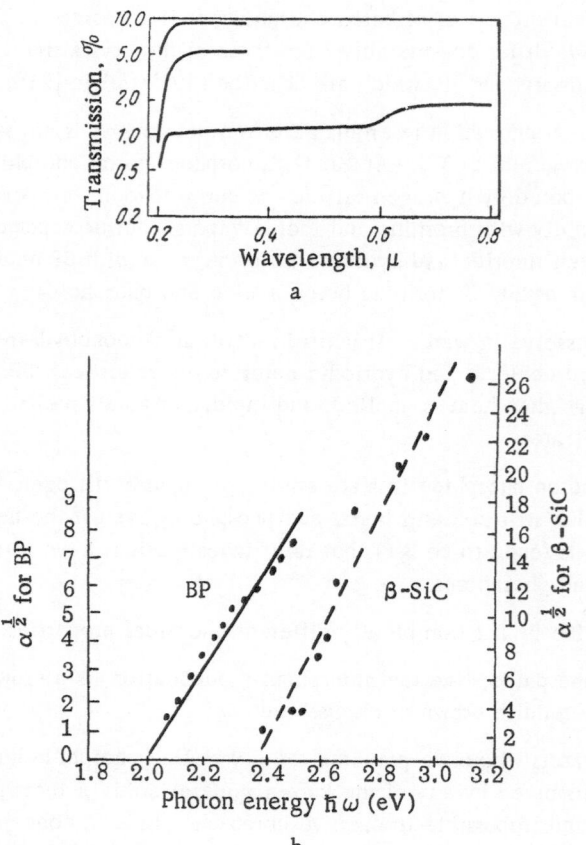

Fig. 41. Absorption spectra of boron phosphide: a) spectrum of BP; b) spectra of BP and β-SiC (for comparison) taken from [335].

The refractive index was determined roughly, using crystals with plane-parallel faces, from the displacement of the image. Its value was between 3.0 and 3.5.

From the measurements of the optical absorption in thin boron phosphide films of various thicknesses, deposited onto a quartz substrate, Stone and Hill [72] deduced that the forbidden band width was 5.9 eV. They obtained this value from the sharp reduction in the transmission coefficient near 0.2 μ (Fig. 41a). The reddish color of the sample was explained by the presence of additional absorption near 0.6 μ which, as assumed by Stone and Hill, was due to the interaction of the radiation with neutral impurity centers. However, recent investigations [334, 335], dealing with the electroluminescence, the photo-emf at a gold contact, and with the optical absorption and reflection of single-crystal samples have indicated that the forbidden band width of boron phosphide at room temperature is close to 2 eV (ranging from 1.97 eV, as determined from the electroluminescence, to 2.02 eV, as determined from the photo-emf at a surface barrier between Au and BP). A typical dependence of the absorption coefficient on the photon energy is shown in Fig. 41b, which compares the data on the absorption in BP and in β-SiC. The linear dependence of $\alpha^{1/2}$

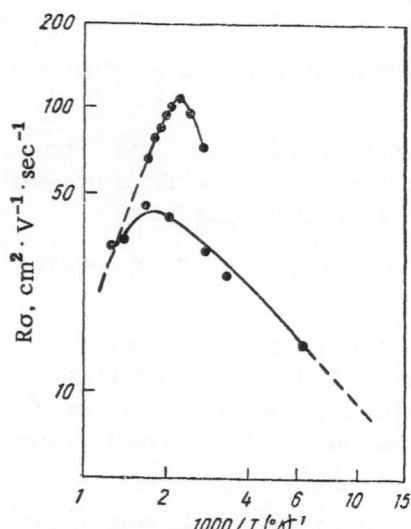

Fig. 42. Temperature dependence of the Hall mobility for boron phosphide.

on $\hbar\omega$ suggests that electron transitions from the valence to the conduction band are indirect and that the band structure of boron phosphide is similar to that of the cubic modification of silicon carbide. By analogy with the cubic silicon carbide, it is assumed that the minimum of the conduction band in BP lies at the edge of the Brillouin zone at the point X. The structure of the reflection spectrum of boron phosphide (maxima at 5.0, 6.9, and 7.9 eV), reported in [335], is associated with direct interband transitions at high-symmetry points in the $\vec{k}$-space.

Measurements of the Hall coefficient were carried out on BP crystals grown from a solution. The sample geometry used was that of the natural crystal faces. Electrical contact was established by means of silvered tungsten or platinum probes. The contact resistance could be reduced by discharging a capacitor through a probe and the sample.

These measurements showed that the crystals were p-type and the carrier density in them was about $(1-5) \cdot 10^{18}$ cm^{-3}; the Hall coefficient did not vary greatly in the temperature range from 900 to 160°K. The form of the resistance curve suggested that at 78°K boron phosphide was still in the region of the total ionization of the impurities. This indicated that $\Delta E_{D,A} \approx 0.05$ eV.

Figure 42 shows the temperature dependence of the Hall mobility for p-type BP samples. The maximum of this curve lies at temperatures higher than room temperature, i.e., at room temperature the scattering on ionized impurities is predominant. In purer p-type samples, the carrier mobility at room temperature should be considerably higher and may possibly reach 300-500 cm$^2 \cdot$ V $\cdot$ sec^{-1}.

In contrast to the crystals grown from solution, the needle-like crystals have n-type conduction. The concentration of impurities in them has been roughly found to be 10^{17} cm^{-3} (assuming that the electron mobility is of the same order as the hole mobility).

Stone and Hill also reported rectification at a point contact with p- and n-type crystals. The thermoelectric power was 300 μV/deg.

Although boron phosphide BP crystals obtained so far still contain a large number of impurities and the data on the properties of this material are purely preliminary, we can conclude that boron phosphide BP is an interesting material from the point of view of its use in various semiconducting devices.

Table 2 gives some data on the properties of boron phosphide.

Applications of Boron Phosphide

The high carrier mobility and the wide forbidden band of BP suggest its use in the preparation of semiconducting devices. As the quality and size of BP crystals are improved, it will be possible to employ them to make power rectifiers, transistors, and possibly, photoelectric devices.

The wide forbidden band and the high chemical stability of BP should favor its application in devices working under extremely difficult conditions (high temperatures and reactive media). The lack of dependence of the Hall coefficient on temperature in a wide range of temperatures (160-900°K), which indicates low ionization energy of the impurities, makes BP an extremely promising semiconducting material.

Boron Arsenide BAs

In contrast to boron phosphide, which was first prepared during the last century, the earliest information on boron arsenide appeared in 1958 [213].

Boron arsenide has been investigated even less than boron phosphide. Judging by published information, this is because it has not yet been possible to obtain this compound in the form of large crystals.

Methods of Synthesizing Boron Arsenide

Only one method of preparing BAs by direct synthesis is known

$$4B + As_4 \rightarrow 4BAs.$$

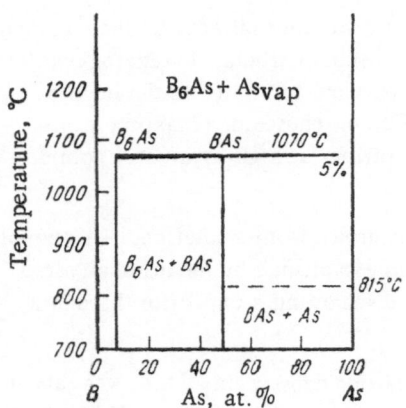

Fig. 43. Phase diagram of the B—As system.

The initial components are amorphous boron and arsenic. The technology of the direct synthesis of boron arsenide is identical with that of boron phosphide, except for different temperature range.

This is because (in full agreement with the relationships governing the properties of analogous compounds, associated with the position of the constituent elements in the periodic system) the binding energy of particles in boron arsenide is less than that in boron phosphide. This should lead to a lower value of the hardness of BAs (compared with BP), a lower melting point, etc., and to a lower temperature at which the dissociation of this compound begins. Therefore, the direct synthesis of boron arsenide should not be carried out at the same high temperatures as that of boron phosphide. An increase in the synthesis temperature above a certain value gives not BAs but the lower arsenide B_6As [213].

The synthesis of boron arsenide is carried out in evacuated, sealed quartz ampoules containing amorphous boron and metallic arsenic. It may be undertaken in a single- or two-temperature furnace. When a two-temperature furnace is used [214], the formation of BAs is observed on heating the hot zone to a temperature not exceeding 700-800°C and the "cold" zone to a temperature which ensures that the vapor pressure of arsenic in the system is not less than 1 atm. Upon increasing the hot-zone temperature to 1000-1100°C and reducing vapor pressure of arsenic, the arsenide B_6As is formed.

Synthesis in single-temperature furnaces is carried out at 760-800°C and it takes several tens of hours. Since at these temperatures a high arsenic vapor pressure may be established, one uses thick-walled (not less than 2 mm) quartz ampoules. If the charge is stoichiometric, the reaction of the formation of BAs is not always completed. The degree of completeness of the reaction (at a given temperature) depends on the vapor pressure of the arsenic in the system. Therefore, it is desirable to have a small excess of arsenic over and above its stoichiometric amount. The product of the direct synthesis, carried out under these conditions, is a black powder. If the ampoule is overheated, arsenic crystals and a yellow powder of B_6As are obtained instead of the black BAs powder.

No methods of separating the synthesis product from unreacted boron have yet been described in the published literature. The unreacted arsenic can be easily removed by vaporizing it and driving it to the cold end of the ampoule. Hence, it follows that, in practice, the charge and conditions of synthesis should ensure the maximum possible yield of BAs in order to reduce the amount of unreacted boron remaining after synthesis.

Medvedeva and Mitkina [321] reported the possibility of preparing boron arsenide BAs in the compacted form. As in the case of boron phosphide, one can use for this purpose compacted boron as the initial material, or one can vibrate the ampoule during synthesis.

Properties of Boron Arsenide

Physicochemical Properties.   The phase diagram of the B—As system has not yet been investigated. Medvedeva and Mitkina [321] studied the products of the chemical interaction of amorphous boron with arsenic over a wide range of concentrations. These products were obtained by saturating amorphous boron powder with arsenic vapor at high temperatures. This investigation made it possible to determine part of the composition — temperature diagram of the B—As system below the solidus line. This diagram is shown in Fig. 43.

The dashed line in Fig. 43 represents the eutectic (degenerate) line. The transformation at 1070°C should be regarded as the peritectic transformation of BAs into B_6As.

This diagram is not an equilibrium diagram because it does not allow for the gaseous-phase pressure.

BAs decomposes when heated in air. In the presence of arsenic vapor, it is stable up to 920°C [213].

Medvedeva and Mitkina [321] mention that the liberation of arsenic from BAs was observed on heating a sealed ampoule (in thermal analysis) to 1070°C. They suggested that the process of decomposition of BAs follows from the equation

$$BAs \rightarrow \tfrac{1}{6}B_6As + \tfrac{5}{6}As,$$

which was confirmed by x-ray diffraction data, and by the quantitative determination of the liberated arsenic.

There are no data on the stability of BAs under the action of various reagents with the exception of a report that BAs dissolves in concentrated boiling nitric acid [214]. The melting point of boron arsenide has not yet been determined. From the relationships governing the properties of $A^{III}B^V$ compounds, we may deduce that this temperature lies between 1800 and 2000°C. Boron arsenide BAs crystallizes in the zinc blende (sphalerite) structure. The lattice period is 4.777 Å. The microhardness of compacted samples of boron arsenide has been determined as 1890 ± 150 kg/mm^2. Table 2 gives some data on the properties of boron arsenide.

Physical Properties. Certain data on the physical properties of boron arsenide, obtained using samples of compacted BAs, are given below [321]:

Resistivity at room temperature,
ρ, $\Omega \cdot cm$ . . . . $4.75 \cdot 10^5$
Thermoelectric power,
α, $\mu V/deg$ 49
Photoconductivity . . . . Not detected

The above values are approximate since the measurements were not conducted on single crystals but on compacted crystals, which were relatively porous lumps of fine boron arsenide crystallites. The dearth of information on boron arsenide prevents our making any predictions about the possible applications of this compound in the preparation of semiconducting devices.

COMPOUNDS BASED ON ALUMINUM

There have been few investigations of aluminum-base $A^{III}B^V$ compounds. Of four compounds of this type, only aluminum antimonide, which has the lowest melting point (1060°C), has been investigated in some detail.

A characteristic feature of these compounds, due to the presence of aluminum, is their corrosion instability in a humid atmosphere and, particularly, in water. This property makes it difficult to investigate and use aluminum compounds. The high melting points of these compounds create difficulties in the preparation of single crystals which are necessary for detailed investigations and applications.

The presence of the very active element aluminum restricts the selection of a suitable material for the containers to be used in the preparation of these compounds. Neither quartz, the silicon in which aluminum reduces easily, nor graphite, with which aluminum forms the carbide Al_4C_3 (these reactions are particularly intense at high temperatures which are necessary to produce Al compounds), can be used as container materials. The most suitable crucible materials for the synthesis of aluminum compounds are oxides of aluminum, magnesium, and beryllium, with which aluminum does not react.

In spite of these difficulties, the present state of the development of the methods of synthesis of semiconductors is sufficiently advanced to make it possible to prepare these interesting compounds, to investigate them, and to find suitable applications.

Aluminum Nitride AlN

Preparation of Aluminum Nitride

Among the compounds considered in this section, aluminum nitride has the highest melting point but it tends to dissociate well before reaching that point. This determines to a considerable degree the method of its preparation. Until recently, AlN was obtained only in the form of powder [225-228], whose semiconducting properties were very difficult to investigate.

Recently, a method has been proposed [229] for preparing AlN, which can simultaneously be used to grow single crystals suitable for some electrical applications. This method is as follows. Above a corundum boat heated to 1500°C and containing molten Al, nitrogen and argon are passed for 24 hours (argon is used as the transport gas). Crystals of aluminum nitride, in the form of plates or needles, 1 × 1 mm and 4 × 0.1 × 0.1 mm, respectively, grow on the walls of a corundum tube in which the boat is placed. The crystals form in that part of the tube where the temperature is about 1450°C and the temperature gradient 15 deg/cm. Tetsuo and Yasaku [229] are of the opinion that the temperature of 1500°C, which they were able to reach under their experimental conditions, is far too low for the preparation of large crystals.

The Growing of Aluminum Nitride Single Crystals

Apart from the method of growing single crystals (which can be carried out simultaneously with the synthesis of AlN) which has just been described, Tetsuo and Yasaku tested a method in which they started with an aluminum powder [229]. They passed nitrogen for 24 hours above a corundum boat containing AlN heated to 1500°C. Rod-shaped crystals grew on the walls of a tube in which the boat was placed but the dimensions of these crystals were considerably smaller than those of the crystals prepared by the method involving simultaneous synthesis of AlN. Tetsuo and Yasaku concluded that again the temperature of 1500°C was too low to prepare large crystals.

It should be possible to prepare large single crystals of AlN by the gas-transport reaction method, as well as by crystallization from solution in a suitable solvent. The special difficulty to be overcome will be the separation of the AlN crystals from an excess of the solvent (in the case of crystallization from the solution) because the aluminum nitride will be dissolved together with the solvent in aqueous solutions of the reagent.

Properties of Aluminum Nitride

Physicochemical Properties. The phase diagram of the aluminum–nitrogen system is completely unknown. All we know is that, apart from the chemical compound AlN, aluminum and nitrogen form aluminum triazide $Al(N_3)_3$, which is a white explosive.

Aluminum nitride is a white or grayish white powder, whose larger crystals are completely colorless and transparent [230]. AlN crystallizes in the wurtzite structure with lattice parameters a = 3.104 kX and c = 4.965 kX [231, 232]. It is very hard (9 units on the mineralogical scale [233]) and its density is slightly more than 3 g/cm^3 [234]. AlN melts at 2200°C (at a pressure of 4 atm [235]) and it sublimates at 1840-1870°C [236, 237].

The chemical properties of AlN have been investigated relatively well. This compound dissolves easily in cold and hot water; it is easily decomposed by water when heated to 100°C, which liberates ammonia NH_3 [230]. Aqueous solutions of acids dissolve AlN to form the corresponding ammonium or aluminum salts [238]. This instability of AlN in aqueous solutions is due to its tendency to hydrolyze. Aqueous solutions of alkalis react much less with AlN but molten alkalis dissolve it. On the other hand, up to 1220°C AlN does not react with carbon, boron, or silicon; it reacts very slowly with phosphorus, oxygen, sulfur, and bromine. Chlorine decomposes aluminum nitride [238].

When heated in air, AlN begins to decompose at 940-950°C [235], although there are some indications that it is more stable — up to 1750°C [227]. The heat of formation is $\Delta H = 60$ kcal/mole [235], and the molecular specific heat of solid AlN in the temperature range 298-900°K may be given by the expression [236]:

$$C_p = 5.47 + 7.80 \cdot 10^{-3}T \text{ cal} \cdot \text{mole}^{-1} \text{ deg}^{-1}.$$

Electrical Properties. Of the electrical properties of aluminum nitride, only the forbidden band width of 3.8 eV (Table 3) is known; this value was obtained by the optical method from the absorption band edge [78].

This value of the forbidden band width is not accurate: some theoretical calculations and the temperature dependence of the conductivity [80] indicate that it should be more than 5 eV.

Other electrical properties of AlN have not yet been investigated because of the difficulty of preparing samples suitable for electrical measurements. The method of preparing single crystals of aluminum nitride

TABLE 3. Principal Properties of Refractory $A^{III}B^V$ Compounds Based on Aluminum

Compound	T_{mp}, °C	Heat of formation, kcal/mole	Structure	Lattice parameter, Å	Density g/cm³	Microhardness, kg/mm²	Hardness on Mohs' scale	Thermal expansion coefficient, deg⁻¹	Forbidden band width, eV	Thermoelectric power, μV/deg	Characteristic temperature, °K	Thermal conductivity, cal·deg⁻¹·cm⁻¹·sec⁻¹
AlN	2200 At p=4 atm [235]	57.7–60 [235,240]	Wurtzite	a = 3.104 kX, c = 4.965 kX, c/a = 1.600 [231, 232]	3.05 [234]	—	9 [233]	—	3.8 [78] > 5 [80]	—	—	—
AlP	1800 [241]	20 [241]	Sphalerite	5.451 [243]	2.42 [244]	—	5.5 [67]	—	2.42 [245]	—	—	—
AlAs	1700 [246]	—	Sphalerite	5.6622 [247]	3.598 [248]	500 ± 20 [249,246]	5	$3.5 \cdot 10^{-6}$ [247]	2.16 [250]	70 [246]	400 [247]	10^{-3} [246]

described by Tetsuo and Yasaku [229] may make it possible to investigate the semiconducting properties of this compound in the near future.

The available data on the principal properties of aluminum nitride are given in Table 3.

Applications of Aluminum Nitride

Aluminum nitride seems a very promising material for crucibles in which other semiconductors, for example, gallium arsenide, can be melted [239]. Aluminum nitride, like other substances of the $A^{III}B^V$ type, has a very narrow range of homogeneity, i.e., it does not dissolve excess aluminum and nitrogen. This property, in conjunction with the high melting point, the high heat of formation, and the high temperature at which dissociation begins (1750°C), is important in the purification of this compound and in the preparation of crucibles from it. It is known that elements of groups III and V of the periodic system are not electrically active impurities in $A^{III}B^V$ compounds. Therefore, aluminum or nitrogen impurities which might enter from a crucible into molten gallium arsenide should not affect the electrical properties of the latter semiconductor. It has been reported [239] that gallium arsenide single crystals prepared in aluminum nitride crucibles have very good electrical properties, such as an electron mobility of 7800 cm² · V⁻¹ · sec⁻¹ at an electron density of (1-3) · 10¹⁵ cm⁻³.

We cannot say anything definite about the use of AlN as a semiconductor since its semiconducting properties have not yet been investigated in full.

Aluminum Phosphide AlP

Until recently, there was no information at all on aluminum phosphide as a semiconductor. Its properties were discussed only in connection with changes in the corrosion and magnetic properties of aluminum alloyed with phosphorus [234] and with a search for a material which can easily liberate the poisonous gas phosphine PH_3 [251].

This lack of information is due to the enormous difficulties in preparing AlP, especially in its pure form. So far, it has been possible to prepare aluminum phosphide only as a powder containing 4-5% impurities.

Preparation of Aluminum Phosphide

Aluminum phosphide AlP was prepared for the first time by an indirect method: fusion of sodium metaphosphate with aluminum powder in a sealed ampoule [252]. The reaction product was a powder which decomposed easily in air. The same type of product was also obtained by other methods [243, 251, 253, 254].

Only recently has an indirect synthesis method, using corundum containers, been applied successfully to prepare aluminum phosphide in the form of small single crystals, on which a number of electrical and optical measurements can be carried out [245]. The procedure was as follows. A corundum boat containing 20 g pure (99.997% Al) aluminum was placed in a corundum tube. The boat was heated to 1200°C, while hydrogen was passed continuously through the tube. At this temperature, the hydrogen was replaced partly or completely with phosphine PH_3, and the reaction chamber temperature was raised to 1500°C. Then the conditions were stabilized and maintained for several hours. The synthesis in the boat produced a compacted mass, covered with a gray-yellow film, under which transparent light-yellow AlP crystals were found, which were produced by the reaction

$$2Al + 2PH_3 \rightarrow 2AlP + 3H_2.$$

These crystals were more than 1 mm^2 in cross section and 0.2 mm thick.

Investigations of the properties of aluminum phosphide, prepared by different methods, showed that the purest crystals were obtained by the method of exchange phosphidization [243], and the method using corundum containers [245].

Obviously, these two methods of synthesis are of great interest. The former gives a pure product suitable as a raw material in growing single crystals by the gas-transport reaction method. The second method, after suitable improvement (for example, by increasing the reaction temperature and the development of suitable conditions) should give large (several millimeters in size) AlP crystals directly at the end of the process.

The Growing of Aluminum Phosphide Single Crystals

No special methods of growing aluminum phosphide single crystals have yet been developed, if we disregard the method (simultaneous with synthesis) of using corundum containers [245]. Apart from this method, there is great interest in the gas-transport reaction methods which have been developed in recent years.

As in the case of aluminum nitride, the crystallization of AlP from solution will be difficult because of the lack of a suitable solvent for AlP and the problem of separating the grown crystals from the excess constituents.

Properties of Aluminum Phosphide

Physicochemical Properties. The phase diagram of the Al—P system has not yet been investigated. It is assumed that, apart from the phosphide with the composition AlP, the system also includes chemical compounds of other compositions: Al_2P_3, Al_3P, Al_5P_3, Al_3P_7. However, the existence of these phosphides is in doubt [244].

Depending on the structure of the component particles and on the impurities present, aluminum phosphide can be green [243] or light yellow [245]. Like all aluminum compounds of this group, AlP is easily hydrolyzed in moist air, liberating phosphine [251]:

$$AlP + 3H_2O \rightarrow Al(OH)_3 + PH_3.$$

AlP decomposes easily under the action of dilute and concentrated acids. However, AlP synthesized from very pure components, even in powder form, has much greater corrosion stability in water and alkalis [243]. Thus, AlP powder placed in water for several weeks was not completely hydrolyzed and x-ray diffraction analysis of this powder showed clearly the AlP structure (in addition to the lines of the decomposition products). This is due to the absence of excess Al or P in such AlP, which are usually present in AlP prepared by direct synthesis methods.

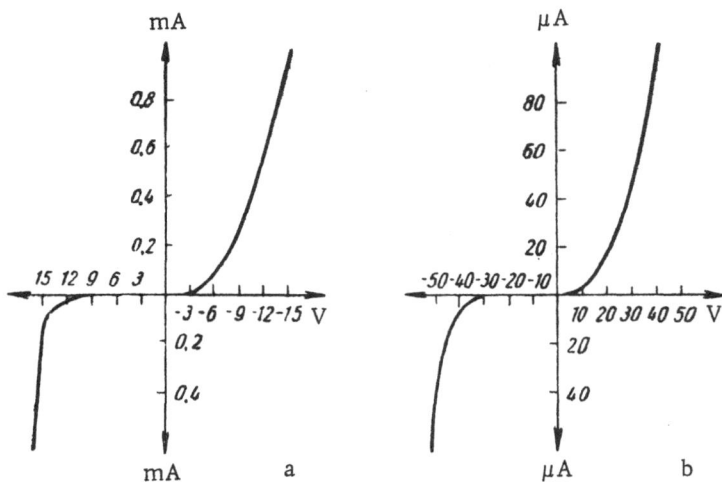

Fig. 44. Current—voltage characteristics of point contacts with
p- and n-type samples of aluminum phosphide.

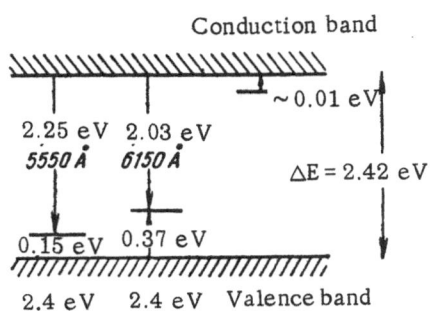

Fig. 45. Proposed level scheme for the
forbidden band of aluminum phosphide.

Aluminum phosphide has a very low vapor pressure at
1000°C, i.e., it practically does not dissociate at this high temperature [243]. Aluminum phosphide crystallizes in the sphalerite
structure with a lattice parameter a = 5.451 Å [243]; its density
is 2.42 g/cm³ [244] and the mineralogical hardness is 5.5 [67].

The melting point has not yet been determined exactly;
it is assumed that it lies near 1800°C [241]; the vapor pressure
at the congruent melting point of AlP reaches fairly high values.

Electrical Properties. Aluminum phosphide is a
typical semiconductor with a forbidden band width of 2.42 eV [245].

The sign of the conduction in AlP may be altered by in-
troducing suitable impurities. To obtain p-type crystals, copper,
zinc, or cadmium is introduced into the aluminum during the
synthesis; to obtain n-type conduction, a small amount of hydrogen sulfide or hydrogen selenide (the doping
impurities are sulfur or selenium) is introduced into the phosphine flowing above the molten aluminum.

Aluminum phosphide exhibits good rectification at a point contact. In the best case, the rectification
coefficient reaches 10^5.

Figure 44 shows the current—voltage characteristics of point contacts with p- and n-type samples of
aluminum phosphide.

Grimmels and Kischio [245] investigated the electrical conductivity, the luminescence of a p-n junction,
and the photoeffect, from which they deduced a scheme of local levels in the forbidden band of aluminum
phosphide (Fig. 45).

The great disadvantage of aluminum phosphide·— and of other aluminum compounds — is its corrosion
instability. However, by analogy with AlSb [255], we may assume that a high degree of purification of the
initial materials before synthesis will give a purer final product, which should markedly increase its chemical
stability. This phenomenon was observed by Addamiano [243], who was able to prepare AlP free of excess
Al and P (see above).

Moreover, it should be possible to prepare solid solutions of AlP with other corrosion-stable phosphides,
for example, with GaP or InP and this, apart from increasing the corrosion stability, should (as in the case of
AlSb [256]) make it possible to obtain conducting materials with new combinations of the principal electrical
and physicochemical properties. Table 3 presents data on some properties of AlP.

Applications of Aluminum Phosphide

The wide forbidden band of AlP suggests that this semiconductor could be used to prepare rectifiers capable of operation at high temperatures. With a forbidden band width of 2.42 eV, luminescence should be observed in the yellow-green part of the spectrum so that AlP is also interesting as an electroluminescent material. Further possible applications of AlP depend on the success of the technique of growing single crystals and protecting them from corrosion, and on the results which will be obtained in further investigations of the semiconducting properties of this compound.

Aluminum Arsenide AlAs

This compound has the lowest melting point among the compounds of aluminum considered in the present book ($T_{mp} \sim 1700°C$). However, the preparation of aluminum arsenide is difficult because of the presence of arsenic. The properties of aluminum arsenide have not been investigated much.

Preparation of Aluminum Arsenide

Aluminum arsenide was obtained first by an indirect method, in the form of a powder [253].

Improved methods of direct synthesis of purified aluminum and arsenic have been used to prepare compacted samples of aluminum arsenide, for example, by the pressing and subsequent firing of tablets made of initially powdered materials [257].

Presnov et al. [246] placed the initial components in a graphite crucible with a cover, sealed the crucible into an evacuated quartz ampoule, and then heated the ampoule slowly in a furnace. A temperature gradient was established along the ampoule; the temperature in the hot part reached 1200°C, while in the cold part it was 600°C (at the beginning of the process). The synthesis took 60 hours. The resultant samples were in the form of dense aluminum arsenide ingots, weighing 10-15 g, which could be used to carry out a number of measurements (see below).

AlAs was also obtained in the form of an ingot by the direct fusion of Al and As in graphite crucibles, placed in evacuated quartz ampoules; the constituents were subjected to vibrational mixing [215]. The latter shortened the synthesis period to 3.5 hr (the charge was $\sim$5 g) and reduced the probability of an explosion of the ampoule (due to the high vapor pressure of arsenic) because the rate of reaction between arsenic and aluminum was accelerated.

In the direct synthesis of AlAs by the fusion of mixed Al and As, the melting point of the component was not reached because above 1400°C the quartz ampoules softened considerably and fractured. Therefore, the reaction between Al and As took place mainly by diffusion in the solid state.

All the AlAs samples, irrespective of the method by which they were prepared, consisted of more than one phase and were not pure. They always contained some excess of aluminum.

Obviously, aluminum arsenide could be obtained by synthesis under pressure, but neither quartz nor graphite, with which Al reacts quite strongly at the required temperatures, are suitable materials from which to make the crucibles or boats to be used in the melting process [258]. The most suitable crucible materials are aluminum and magnesium oxides and a number of other refractory oxides.

The Growing of Aluminum Arsenide Single Crystals

The methods of growing AlAs single crystals have not yet been developed. However, it is clear which methods would be most suitable for this compound. First of all, the gas-transport reaction method, with halogens as the transport gases, would be convenient. However, this method can hardly give very large single crystals but it should produce needles and plates of up to several millimeters.

The method of pulling single crystals from a nonstoichiometric melt (with a large excess of Al) by the Czochralski technique, as used for InP (see below), should be interesting.

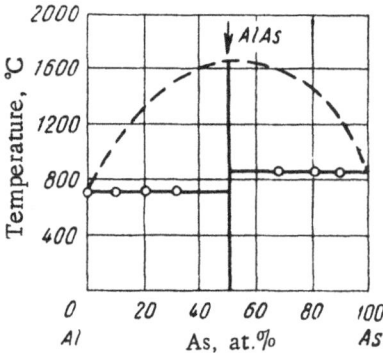

Fig. 46. Phase diagram of the Al—As system.

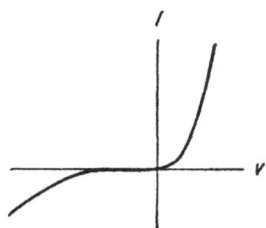

Fig. 47. Current—voltage characteristic of a point contact with aluminum arsenide.

The method of crystallization from solution (for example, in aluminum) involves the difficulty of separating the easily corroded aluminum arsenide from a solvent (Al) with a relatively high melting point.

Properties of Aluminum Arsenide

Physicochemical Properties. The phase diagram of the Al—As system has been constructed from thermal and metallographic analyses. It is given in Fig. 46 [257].

It is evident from this figure that the Al—As system forms one chemical compound at the equiatomic ratio of the constituents. It is reported that AlAs melts without decomposition at some temperature greater than 1600°C. However, this seems doubtful because all the analogs of aluminum arsenide (indium and gallium arsenides and phosphides) give off the volatile component on melting. Aluminum arsenide oxidizes when heated, does not react with bromine and iodine, and is a very strong reducing agent [248].

In the powdered form, aluminum arsenide is brown, but in the form of ingots (obtained by melting in graphite), it is gray with a metallic luster. AlAs crystallizes in the sphalerite structure with a parameter a = 5.6622 Å [247]. The density is 3.598 g/cm^3 [248], and the microhardness is ~ 500 kg/mm^2 [249, 246].

Aluminum arsenide is very unstable in a humid atmosphere. It rapidly corrodes, undergoing the hydrolysis:

$$AlAs + 3H_2O \rightarrow AsH_3 + Al(OH)_3.$$

In water, and even more so in dilute acids, AlAs decomposes very rapidly, giving off arsine (AsH$_3$), a strongly poisonous gas; when AlAs decomposes in water, one of its reaction products forms the acid H$_3$AsO$_3$ [248]. The process of hydrolysis of aluminum arsenide is strongly aided by the presence of excess aluminum, which can be seen clearly in microsections of AlAs ingots. Probably a large potential difference appears between these aluminum occlusions and the matrix, and this potential difference accelerates strongly the rate of decomposition of the compound in a humid atmosphere or in aqueous media (a similar effect is also observed in aluminum antimonide). We may assume that the removal of these aluminum occlusions and an increase of the purity of the material would improve its corrosion stability.

Moreover, the corrosion instability of aluminum arsenide can be reduced considerably by adding stabilizing components, which form solid solutions with aluminum arsenide. Such components might be, for example, indium arsenide InAs [259], and gallium arsenide GaAs [221], with which aluminum arsenide forms substitutional solid solutions. The papers just cited [259, 221] mention that such additions do reduce the corrosion of AlAs. The synthesis of the solid solutions can be carried out by the direct melting of the elements involved in an evacuated chamber and the resulting solutions can be homogenized by the recently developed method of self-fed melt [260].

Electrical Properties. Aluminum arsenide is a typical semiconductor with a wide forbidden band. According to Herman [250], the forbidden band width is 2.16 eV (it was determined by an optical method).

The carrier mobility in aluminum arsenide is not known. In Suchet's theoretical paper on semiconductors [261], it is suggested that the electron mobility may be high in AlAs. Suchet calculated, bearing in mind the possibility of ionicity of A^{III}B^{V}-type compounds, that the electron mobility in AlAs should reach 70,000 cm$^2 \cdot$ V$^{-1} \cdot$ sec^{-1}. However, since the properties, particularly the mobility, change considerably through a semiconducting series, we can hardly expect AlAs to have this mobility. The electron mobility in aluminum arsenide cannot be higher than 200-500 $cm^2 \cdot V^{-1} \cdot sec^{-1}$. This value may be increased by dissolving, in AlAs, similar compounds with a high carrier mobility, for example, indium and gallium arsenides mentioned earlier in connection with increasing the corrosion stability (the electron mobilities in InAs and GaAs are 30,000 and 8,000 $cm^2 \cdot V^{-1} \cdot sec^{-1}$, respectively).

The thermoelectric power of aluminum arsenide, measured using dense polycrystalline samples, is ~ 70 $\mu V/deg$ [246]. Presnov et al. [246] also reported the rectification at a point contact. The nature of the current—voltage characteristic can be seen in Fig. 47.

Table 3 lists the principal properties of AlAs.

Applications of Aluminum Arsenide

One cannot expect any wide use of AlAs in the immediate future. Only a careful investigation of the electrical properties of aluminum arsenide may show how promising is this compound, since its instability in air would complicate considerably the manufacture of devices made from it.

COMPOUNDS BASED ON GALLIUM

$A^{III}B^{V}$-type compounds based on gallium have one thing in common: the low melting point of gallium (29.8°C) facilitates the selection of a suitable method for the synthesis of these materials.

On the other hand, the ability of molten gallium to dissolve quite vigorously the silicon present in quartz containers, which are usually employed for melting, makes it necessary to take preventive measures.

Apart from the compounds considered separately below, gallium-base compounds include gallium antimonide and arsenide. These two compounds have been investigated very thoroughly; this is particulary true of gallium arsenide, which at present is one of the most important semiconductors because it is used to prepare such basically new devices as tunnel diodes and lasers. Gallium antimonide is also used for the same purposes.

These two compounds are not considered in the present book. Gallium antimonide cannot be regarded as refractory (it melts at 712°C), while the properties of gallium arsenide can be found in many papers and reviews, as well as in a number of extensive monographs [24, 36, 79].

Gallium phosphide and, particularly, gallium nitride have been investigated much less. The remarkable properties of gallium phosphide (which are considered below) make it possible to produce new semiconducting devices with interesting characteristics. In particular, its electroluminescent properties are very promising. However, GaP has not been investigated much because of the considerable difficulties encountered in preparing this compound.

Gallium Nitride GaN

Preparation of Gallium Nitride

So far, gallium nitride has been prepared only by indirect methods. This is due to the very low rates of reaction of gallium with gaseous nitrogen. The use of indirect synthesis methods produces gallium nitride only in the form of fine-grained powders.

The majority of methods of preparing GaN reduces to the treatment of molten gallium, heated to various temperatures (from 350 to 1200°C), with ammonia [29, 262-264, 227, 265, 266]. Like indium nitride (see below), gallium nitride has also been obtained by the thermal decomposition of a double salt $(NH_4)GaFe_6$ in an ammonia atmosphere (at 900°C) [267].

Attempts have been made [228] to use gallium arsenide and phosphide, previously prepared in an exchange reaction with ammonia, in accordance with the equations

$$GaP + NH_3 \rightarrow GaN + PH_3$$

TABLE 4. Principal Properties of Refractory $A^{III}_{B}V$-type Compounds Based on Gallium

Compound	T_{mp}, °C	Heat of formation, kcal/mole	Structure	Lattice parameter, Å	Density, g/cm³	Microhardness, kg/mm²	Thermal expansion coefficient, deg^{-1}	Forbidden band width, eV (at 20°C)	Electron mobility, cm²·V^{-1}·sec^{-1} (at 20°C)	Hole mobility, cm²·V^{-1}·sec^{-1} (at 20°C)	Refractive index	Permittivity
GaN	~1500 [67]	24.9 [267]	Wurtzite	a = 3.180, c = 5.160 [29]	6.1 [29]	—	—	3.25 [270]	—	—	—	—
GaP	1467 ± 3 At p = 35 ± 10 atm [293]	14.35 ± 08 [293]	Sphalerite	a = 5.45 [274, 272, 243, 300]	4.10 [292]	940 ± 35 [249]	5.3 · 10^{-6} [295]	2.25 [296]	110–130 [298, 299, 279]	150 [279]	3.37 [296], 2.9 [297]	8.5 ± 0.2 At λ = 1.12 μ, 10.2 At > 40 μ [86]

and

$$GaAs + NH_3 \rightarrow GaN + AsH_3.$$

However, this method involves a considerable expenditure of effort in preparing the pure intermediate products (GaP and GaAs). Moreover, in this case, the reaction cannot be completed, even at high temperatures (> 1000°C).

The synthesis process investigated most is the reaction [268]

$$Ga_2O_3 + 2NH_3 \rightarrow 2GaN + 3H_2O.$$

When ammonia is passed over gallium oxide heated to various temperatures (480–1090°C), a powder is obtained whose color varies from light yellow to light gray as the synthesis temperature is increased. We may conclude that the indirect synthesis methods will continue to be the major ones in the preparation of gallium arsenide.

Growing of Gallium Nitride Single Crystals

GaN single crystals have not yet been prepared probably because of the high melting point of this compound and its probable dissociation at elevated temperatures. The vapor pressure at the congruent melting point of gallium nitride is certainly not less than the corresponding vapor pressure of gallium phosphide (35 ± 10 atm).

For these reasons, the growing of single crystals from stoichiometric melts is hardly possible.

It is more likely that GaN can be grown from solutions. However, the problem of finding a suitable solvent meets with considerable difficulties (see the first part of the book). The use of easily melted gallium as a solvent, which has the advantage of being one of the constituents of the compound, depends on the solubility of GaN in Ga at the maximum possible temperatures under actual laboratory conditions (for example, quartz ampoules are used at 1300–1350°C). Judging by the phase diagrams of similar systems (Ga–P, In–P, etc.), the appropriate solubility of GaN should not be greater than 1–2% and this value is too small to hope for large single crystals.

A more promising method of producing at least small (up to several millimeters) single crystals of GaN is the gas-transport reaction method (cf. the first part of the book).

Properties of Gallium Nitride

Physicochemical Properties.   The phase diagram of the Ga−N system has not been investigated at all.   It is only known that Ga and N form the chemical compound GaN (with the equiatomic ratio of its constituents), and gallium triazide $Ga(N_3)_3$, which can be obtained by adding HN_3 to an ether solution of gallium hydride frozen by cooling with liquid nitrogen.

Gallium nitride prepared by the methods described above is a fine-grained powder, whose color varies somewhat with the temperature at which it has been prepared (the color ranges from light yellow to light gray).

GaN crystallizes in the wurtzite structure with the lattice parameters a = 3.180 Å and c = 5.160 Å; the density is 6.1 g/cm^3 [29].   The heat of formation of this substance from its elements amounts to $\Delta H = 24.9$ kcal/mole.   GaN is chemically stable, and it does not oxidize or decompose in air, in water, or in dilute acids and alkalis.   It dissolves very slowly in hot concentrated sulfuric acid and in an aqueous solution of sodium hydroxide.   GaN does not react with hydrogen, and when it is heated in an oxygen atmosphere, gallium oxide Ga_2O_3 begins to form only at 900°C [264, 230, 269].

Until recently it has been assumed that GaN does not decompose when heated up to 1250°C [270].   A more recent investigation has shown that the thermal dissociation of GaN becomes already marked at 616°C [268].

Thus, the use of GaN and its alloying (activation) are limited by the temperature at which dissociation begins (~600°C), but even at higher temperatures the dissociation is not very strong.

Data on some of the properties of GaN are given in Table 4.

Electrical Properties.   The electrical properties of GaN have not been investigated at all.   All that is known is that its electrical resistivity, measured using slabs pressed from powder, is very high ($\sim 4 \cdot 10^8$ $\Omega \cdot cm$) [270].

This has been confirmed by Ormont [73].   The high resistivity is also in agreement with the optically determined forbidden band width of 3.25 eV [270]; Ormont reports a calculated value of 3.6 eV [73].

Gallium nitride, like other compounds of its group having a wide forbidden band, exhibits interesting luminescent properties.   Two intensity maxima — at 3200 and 5200 Å — were found in an investigation of the cathodoluminescence of GaN powders, prepared at various temperatures [268].

Applications of Gallium Nitride

It is difficult at present to speak of the applications of GaN but it is evident that this semiconductor may prove useful in the preparation of high-temperature semiconducting devices.

Gallium Phosphide GaP

Gallium phosphide has become one of the more important semiconducting materials.   Its high melting point and high thermal stability have made it possible to produce diodes working at temperatures above 500°C [271].   The electroluminescent properties of GaP are also very interesting and their utilization may give rise to a number of new devices.

Preparation of Gallium Phosphide

Gallium phosphide was first prepared by an indirect method, by passing phosphorus vapor in a stream of hydrogen over gallium hydroxide heated to 500°C [272]:

$$Ga(OH)_3 + P + \frac{3}{2}H_2 \rightarrow GaP + 3H_2O.$$

GaP may be prepared also by an exchange reaction of the phosphiding type [273]:

$$2GaCl + 2P \text{ (or } PH_3) \rightarrow 2GaP + Cl_2.$$

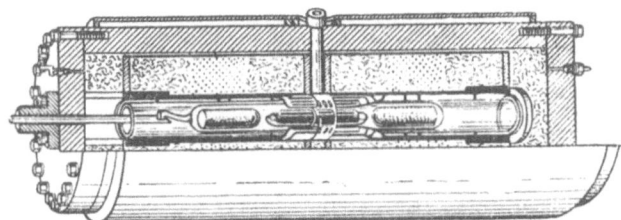

Fig. 48. Apparatus used to melt gallium phosphide under pressure.

Relatively recently gallium phosphide has been prepared by passing, for 7 hours, phosphine (PH_3) diluted with argon or hydrogen over gallium oxide heated to 900-950°C [274]:

$$Ga_2O_3 + 2PH_3 \rightarrow 2GaP + 3H_2O.$$

Prolonged heating (for several days) at 800°C of zinc phosphide previously prepared and mixed with metallic gallium gives an alloy of gallium phosphide with an excess of zinc and zinc phosphide [243]:

$$Zn_3P_2 + 2Ga \rightarrow 2GaP + 3Zn.$$

The excess zinc and zinc phosphide may be easily expelled by heating the gallium phosphide crystals to 1000-1100°C. The gallium phospide obtained by these methods is a fine powder, which can be used as the raw material for the growing of single crystals, for example, by the gas-transport method.

Gershenzon and Mikulyak [275] synthesized and simultaneously grew small gallium phosphide single crystals in the form of whiskers and needles. In this method, gallium monoxide Ga_2O, kept at 800-1000°C, reacted in the vapor phase with phosphorus, and single crystals of GaP grew on the walls of a tubular quartz reaction chamber.

The use of indirect methods avoids the difficulties of the direct methods, which are dealt with below. For example, the use of indirect methods makes it possible to reduce the temperature of the synthesis. However, the participation of auxiliary substances (hydrogen, argon, various chemical reagents) in the reactions increases the degree of contamination of the final products.

The preparation of gallium phosphide by direct methods is difficult for two main reasons: 1) the difficulty of selecting a suitable material for a reaction chamber which would withstand high temperatures and pressures; 2) the danger of the contamination of the product with the material used to make the reaction chamber and with impurities contained in this material.

Gallium phosphide was prepared for the first time by the direct fusion of its elements — taken in the stoichiometric ratio — by means of vibrational mixing of the reacting materials in a single-temperature furnace [215].

However, gallium phosphide prepared in this way is a very porous fine-grained cake. It is formed at high temperatures (1450°C)* close to its melting point, in contact with a quartz container, and therefore it contains considerable amounts of impurities.

Analysis of the published data shows that the methods of preparing gallium phosphide by direct synthesis are developing in two directions: a) melting under pressure (with counter pressure from an inert gas); b) preparation of crystals from a molten solution.

The former approach, which involves the use of complex apparatus, high temperatures and pressures, is very difficult from the technical point of view. For example, some workers [258, 276] have synthesized GaP in a sealed quartz ampoule, which was placed in a stainless steel tube. The quartz ampoule was subjected to an external pressure (counter pressure) of an inert gas (~20 atm). The melting occurred as the ampoule was moved through a ring-shaped high-frequency heater placed, together with additional platinum resistance heaters, inside the stainless steel tube. Figure 48 illustrates such an apparatus.

Fine-grained ingots of GaP (several centimeters long) prepared by this method were found to be contaminated by the material of the boat (quartz or graphite) in which the melting was carried out and it was

---

*In this case, the walls of the quartz ampoule were very thick (3-3.5 mm) and, moreover, temperatures higher than 1300°C were applied for a relatively short time (10-20 min). Under such conditions, the quartz did not have time to soften markedly.

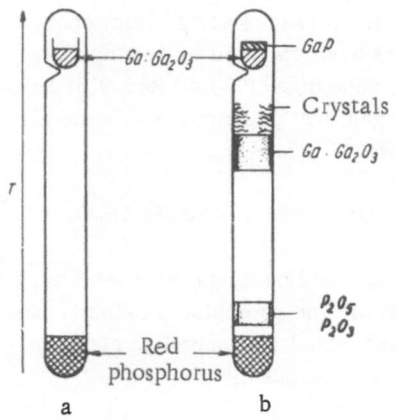

Fig. 49. Quartz container for the preparation of gallium phosphide needles: a) before a run; b) after a run.

therefore necessary to subject these ingots to vertical floating-zone melting in order to purify them and to grow single crystals (see below).

Another method of synthesizing gallium phosphide is to prepare it from a molten solution. Technically this method is the simplest although large single crystals cannot be obtained by using it. However, to prepare many semiconducting devices one needs only very small crystals and therefore the attempts to grow large single crystals are not always justified.

The methods of synthesizing refractory semiconductors from molten solutions are widely used at present. They are technically very simple and can be employed in any laboratory (cf. Chapter I).

The preparation of gallium phosphide by such a method was reported first by Wolff, Keck, and Broder [101]. An excess of gallium was dissolved in hydrochloric acid. Refractory solid solutions can be obtained by the same method [277, 221].

A variant of the synthesis of gallium phosphide from a molten solution is the synthesis in a two-temperature horizontal furnace, followed by directional crystallization [84].

This method gives quite large polycrystalline ingots (up to several centimeters). The maintenance of a fixed temperature at the "cold" end of the ampoule makes it possible to carry out the process under a controlled phosphorus vapor pressure not exceeding the strength of the ampoule.

The Growing of Gallium Phosphide Single Crystals

Most of the known methods of preparing gallium phosphide single crystals are directly associated with the synthesis techniques just described, so that the process of growing single crystals of gallium phosphide is simultaneous with the synthesis of the compound. However, while the synthesis involves a chemical reaction between gallium and phosphorus, which ends when the reaction is complete, the preparation of single crystals involves, almost always, a cooling stage, i.e., the crystallization of the product from the melt or from the gaseous phase.

We shall take first the possibility of cutting very small single-crystal samples from polycrystalline ingots. The obvious disadvantages of this method are the substantial loss of material in the cutting and grinding of the samples, and the indeterminate orientation of "single crystals" so obtained. It should be mentioned also that the preparation of a sufficiently coarse-grained ingot of gallium phosphide is itself very difficult. These "single crystals" are cut mainly from ingots prepared by melting under pressure [258, 278]. A number of studies have been made of such single-crystal samples [279-283, 86]. In these investigations, the use was made of polycrystalline ingots prepared by the directional cooling of concentrated GaP solutions in Ga in a two-temperature furnace [84].

The most promising method of preparing single crystals is the floating-zone technique, which has already been used successfully [258, 278]. Ingots, prepared in a high-pressure apparatus (cf. Fig. 48), were cut into slabs which were subjected to vertical floating-zone melting without the use of a high-pressure chamber. The quartz tubes did not explode because they had thick walls.

This method produced ingots (5-6 mm in diameter, and up to 100 mm long) from which large single crystals of GaP (up to 20 mm long and 5 mm in diameter) were cut. Some investigations have been carried out on such samples and the results are given below [281].

The complexity of the apparatus used for floating-zone melting and the difficulty of preparing rods suitable for melting restrict the use of this method to grow single crystals. Moreover, the melting process itself is impeded by the presence of a large amount of gallium phosphide in the gaseous phase.

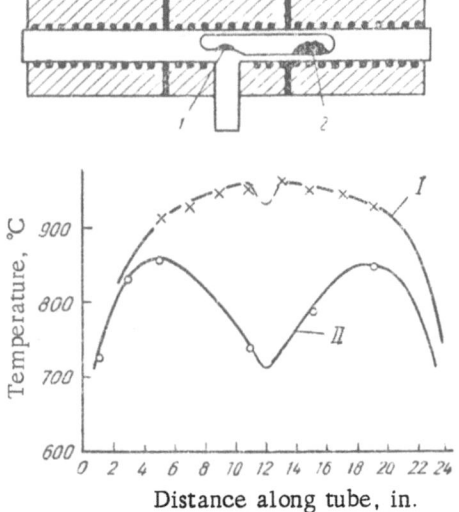

Distance along tube, in.

Fig. 50. Schematic representation of the apparatus and temperature distribution for the growth of gallium phosphide crystals using iodine as the transport gas: 1) crystals; 2) powder; I) at the beginning of a run; II) at the end of a run.

The gaseous product is deposited on the walls of a quartz tube and this makes it difficult to observe the behavior of the molten zone. Nevertheless, this method will probably be used more widely in the technology of gallium phosphide.

The group of methods for growing single crystals of GaP from the gaseous phase is of immense interest. When gallium phosphide is melted in an ampoule, there is usually a temperature gradient. GaP sublimates from the hot zone and, under favorable conditions, small single crystals, which are quite suitable for some investigations, grow in the cold parts of the ampoule [289].

A different method of some interest has been described by Gershenzon and Mikulyak [275]. It is shown in Fig. 49.

The upper end of a tube, which is kept at 975-1060°C and contains a mixture of Ga and Ga_2O_3 in the ratio 4:1, provides the whole tube with the volatile gallium suboxide Ga_2O. The reaction, in the gaseous phase, of gallium suboxide with phosphorus vapor,

$$Ga_2O + P \rightarrow GaP + P_2O_5 \, (P_2O_3)$$

produces needles, whiskers, and filaments of gallium phosphide on the cold walls of the ampoule.

By this method, one can obtain needles up to 20 mm long and up to 100-200 μ thick, as well as very small lens-shape platelets, 0.1-2 mm long, 1-2 μ thick, and 30-80 μ wide. The same method can be used to dope GaP (simultaneously with its growth) with, for example, sulfur. The advantages of the method are: the relative simplicity and the production of needles of much greater mechanical strength than the plate-shaped crystals; this is ascribed to the more perfect (dislocation-free) structure of the needles. The more perfect structure is indicated by the fact that the minority carrier lifetime in needles is longer than that in the other modifications of single crystals. Alloyed diodes, prepared from high-resistivity needles, had a breakdown voltage of 135 V (for plates grown from the melt this voltage was ~40 V). The great disadvantages of this method are its inability to produce large crystals and its very low yield.

The method of gas-transport reactions, already described (cf. the first part of the book), is of considerable interest. In this case, one uses a transport agent which regenerates its transport ability continuously and produces cyclic motion in the chamber throughout the whole process.

An example of the use of gas-transport reactions in the growing of gallium phosphide single crystals is described by Antell [106].

Gallium phosphide powder, previously prepared by synthesis, and some iodine, the latter acting as the transport agent, were placed in an evacuated sealed ampoule. At the hot end of the ampoule, the reaction

$$3GaP + 3I \rightleftharpoons 3GaI + 3P$$

"shifts" to the right and the gaseous products of this reaction, which occupy the whole volume of the ampoule, produce gallium phosphide in the form of needles

$$3GaI + 2P \rightarrow GaI_3 + 2GaP.$$

Figure 50 shows schematically the apparatus and temperature conditions used to prepare gallium phosphide crystals employing iodine as the transport gas.

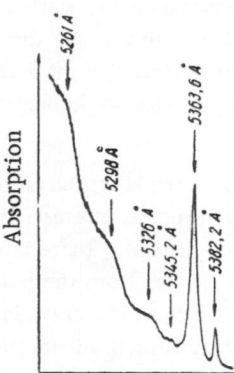

Fig. 51. Microphotogram of the absorption spectrum of a gallium phosphide crystal at 77°K.

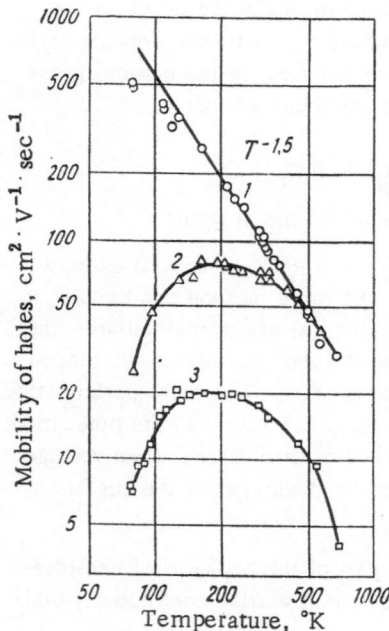

Fig. 52. Temperature dependence of the mobility of holes in GaP: 1, 2, 3) sample numbers (impurity concentration increases with the sample number).

By bringing a metal rod into contact with the ampoule, local heat transfer may be established and the nucleation of single crystals of GaP may be observed.

If a suitable temperature is selected for the central zone, only several crystals are left in the cold zone (window) after the local cooling is stopped. Then the central zone of the furnace is cooled slowly (for several days). As a result, needle-shaped gallium phosphide crystals grow in the cold zone of the ampoule. It is found that the iodine does not dissolve in the GaP at all and the resistivity of crystals obtained in this way is very high ($5 \cdot 10^9$ $\Omega \cdot$ cm). This method is not yet capable of producing GaP single crystals longer than 5-10 mm. It is very difficult to separate the grown crystals from the quartz ampoule walls and they frequently break.

The growing of single crystals from a dilute solution of phosphorus in gallium is an exceptionally important method of growing single crystals of GaP (and of other semiconductors). This method is similar to the synthesis of GaP described earlier. As in the latter method, a charge placed in an evacuated quartz ampoule is heated to temperatures somewhat greater than the liquidus temperature and then the charge is cooled slowly in accordance with a certain program. Favorable conditions are established in the molten solution for the growing of relatively large single-crystal plates (up to 1 cm^2 in area and more than several tenths of a millimeter thick).

Obviously, this method is the simplest of all those discussed so far. Its great advantage lies in the fact that single-crystal plates are always oriented in the same way [along a (111) plane], have relatively perfect surfaces, and can be very pure due to the relatively low synthesis temperatures (compared with the melting point of stoichiometric GaP, which is 1467°C) and the absence of direct contact between the gallium phosphide and the quartz of which the crucible is made.

The possibility of the simultaneous synthesis of the compound and the growing of single crystals of it, together with the possibility of avoiding the laborious and expensive operations of cutting and grinding, makes this method one of the most convenient for the preparation of GaP single crystals. It has a further advantage that growing GaP crystals are easily doped during melting. Thus, p-type crystals with a hole density of $2 \cdot 10^{17}$ cm^{-3} at room temperature were obtained by adding zinc [284].

The method of growing GaP single crystals from a dilute solution of phosphorus in gallium has some disadvantages as well. Obviously, the growth of single crystals from such a solution in a closed chamber, as described above, takes place under conditions of continuous variation of the phosphorus concentration in the Ga and this, as is known from the theory of crystal growth, makes the growth processes more difficult and distorts them. Moreover, recent investigations [285] have demonstrated that GaP plates prepared by this method are not, strictly speaking, single crystals. It is found that twinning takes place in planes parallel to the most developed surface of the plates (111). Such plates consist at best of two single-crystal plates separated by one boundary. Nevertheless, twinned gallium phosphide single crystals can be used in technical applications.

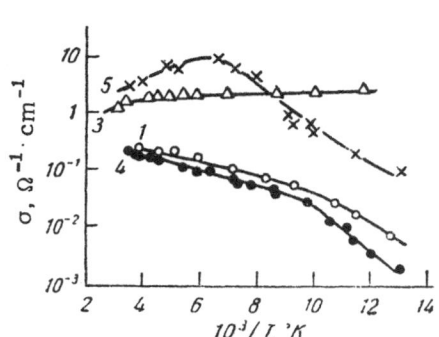

Fig. 53. Temperature dependence of the electrical conductivity of GaP (numbers by the curves denote sample numbers; sample No. 5 is n-type, the rest are p-type).

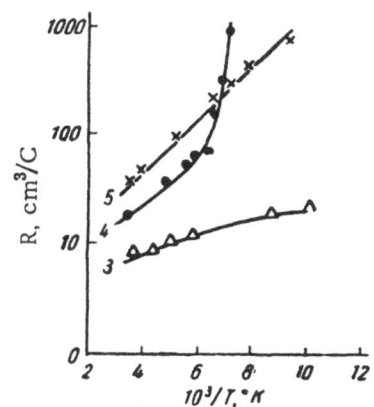

Fig. 54. Temperature dependence of the Hall coefficient of GaP (the numbers by the curves denote sample numbers).

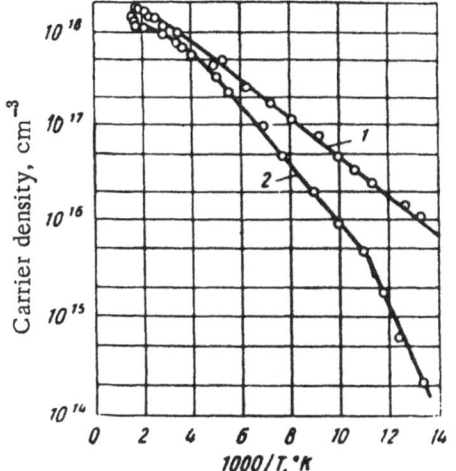

Fig. 55. Temperature dependence of the carrier density in p- and n-type GaP: 1) Zn acceptors ($E_A = 0.041$ eV); 2) S donors ($E_D = 0.123$ eV).

The method described above is now widely used to prepare GaP crystals for investigations of their properties [101, 277, 281, 286-290, 336].

Properties of Gallium Phosphide

Physicochemical Properties. Gallium phosphide crystals are transparent and orange. Like other members of the $A^{III}B^V$ group, GaP exhibits anisotropy along the [111] direction, which is manifested by the difference between the surface properties of crystals on the gallium side [plane (111)] and on the phosphorus side [plane ($\bar{1}\bar{1}\bar{1}$)].

This phenomenon appears very clearly when suitably oriented single crystals are etched [291]. GaP, which crystallizes in the zinc blende structure, has a lattice period $a = 5.45$ Å, and its density, calculated from the x-ray data, is 4.15 g/cm³, which is in good agreement with the experimental result, 4.10 g/cm³ [292].

Gallium phosphide exhibits high chemical stability. It does not oxidize in air up to 750°C and in vacuum it begins to dissociate at 1000°C [84].

The most reliable data on the congruent melting temperature of gallium phosphide and on its dissociation vapor pressure at this temperature have been given by Richman [293]: they are, respectively, 1467 ± 3°C and 35 ± 10 atm.

The phase diagram of the Ga—P system has not yet been published. All that is known is that the system includes one chemical compound with the equivalent proportions of gallium and phosphorus, which forms a "degenerate eutectic." The solubility of gallium phosphide in metal solvents (Ga, In, Sn, Pb, Bi) has been investigated at 1000-1100°C [294]. However, the details of this interesting investigation have not yet been published.

The heat of formation of GaP has been calculated to be 14.35 ± 0.8 kcal/mole; its microhardness has been measured as 940 ± 35 kg/mm² [249]; and the thermal expansion coefficient is known: $5.3 \cdot 10^{-6}$ deg⁻¹ [295].

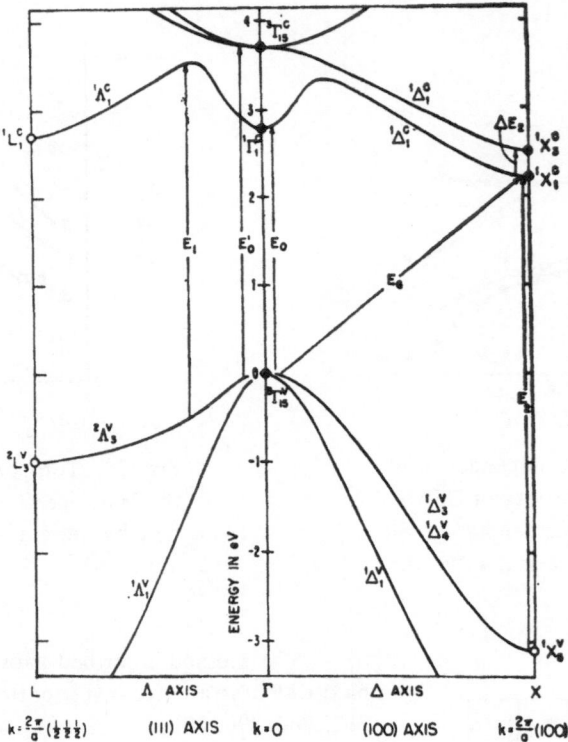

Fig. 56. Band structure of GaP. The states are labeled using the notation for the irreducible representations of the single group of the zinc blende lattice. The left-hand superscript appended to each symbol gives the orbital degeneracy of the state. The arrows represent the transitions observed in optical experiments. The five energy levels marked by solid circles were studied in pressure experiments [342].

The solubility of impurities in gallium phosphide has not been studied much. A recent paper [284] reports an examination of the solubility and diffusion of zinc in GaP. The diffusion was carried out from the gaseous phase at temperatures of the GaP sample of 700-1300°C. The maximum solubility of zinc, $4 \cdot 10^{20}$ cm^{-3}, was observed at 1200°C. The process of the diffusion of Zn in GaP obeys the law $D = 1.0 \cdot \exp(-2.1/kT)$, where D is the diffusion coefficient in cm^2/sec.

Electrical and Optical Properties. The electrical and optical properties of GaP have not yet been investigated sufficiently fully; however, there are grounds for predicting that this compound will soon become technically important because of a very interesting combination of properties.

The influence of hydrostatic pressure on the following optical properties of GaP has been investigated at room temperature:

1)   the fundamental absorption edge in the photon energy range from 2.2 to 2.7 eV;

2)   the infrared absorption band of n-type GaP in the range 0.3-0.5 eV;

3)   the maxima in the reflection spectrum at 2.8 and 3.7 eV;

4)   the recombination radiation of p-n junctions, biased in the forward direction, in the range 1.7-2.3 eV. This made it possible to determine the parameters of the band structure of GaP [342].

It was found that:

a)   the forbidden band width for indirect transitions at 300°K depends on pressure (here and in subsequent formulas, the pressure is given in megabars):

$$E_G \, (\Gamma_{15}^V \rightarrow X_1^C) = 2.22 - 1.1 \; P \; (eV);$$

b)   the temperature dependence of the forbidden band width is:

$$E_G \, (\Gamma_{15}^V \rightarrow X_1^C) = 2.22 - 5.2 \cdot 10^{-4} \; T \; (eV);$$

c)   the forbidden band width for direct transitions at 300°K is:

$$E_O \, (\Gamma_{15}^V \rightarrow \Gamma_1^C) = 2.78 + 10.7 \; P \; (eV);$$

d)   the temperature dependence of the forbidden band width for direct transition is:

$$E_O \, (\Gamma_{15}^V \rightarrow \Gamma_1^C) = 2.78 - 4.6 \cdot 10^{-4} \; T \; (eV);$$

e)   the energy separation between the valence band and the doubly degenerate conduction band at k = 0 at 300°K is:

$$E_O' \, (\Gamma_{15}^V \rightarrow \Gamma_{15}^C) = 3.7 + 5.8 \; P \; (eV);$$

f)   the energy separation between the two conduction bands at the edge of a Brillouin zone along the < 100 > direction at 300°K is:

$$\Delta E_2 \, (X_1^C \rightarrow X_3^C) = 0.3 + P \; (eV).$$

All these transitions are given in Fig. 56.

In contrast to many other $A^{III}B^V$ compounds, the fundamental absorption band edge of gallium phosphide (Fig. 51) lies in the visible part of the spectrum [286]. Calculations of the refractive index have given a value of 3.37 [296], which differs substantially from the experimentally determined value 2.9 [297], found from the reflectivity.

The data on the carrier mobility in GaP are contradictory because of the differences in the quality of samples. The highest mobility — 150 $cm^2 \cdot V^{-1} \cdot sec^{-1}$ (at room temperature) — has been obtained in single-crystal samples, measuring $2.4 \times 1.9 \times 6.9$ mm and cut from polycrystalline ingots. The temperature dependences of the hole mobility are given in Fig. 52 [281], in the region where holes are scattered on the lattice vibrations and the hole mobility is proportional to $T^{-1.5}$. The GaP samples used to obtain the curves in Fig. 52 were strongly compensated and had a hole density of about 10^{18} cm^{-3}. Gershenzon and Mikulyak [281] suggested that hole mobilities of up to 250 $cm^2 \cdot V^{-1} \cdot sec^{-1}$ might be obtained at room temperature in pure samples.

Figures 53 and 54 also give the temperature dependences of the electrical conductivity and of the Hall coefficient of GaP in the temperature range 77-300°K. It has been established that the ionization energy of impurity centers (donors) amounts to $7.5 \cdot 10^{-2}$ eV.

Figure 55 shows the temperature dependence of the carrier density in n- and p-type GaP. The ionization energies of donors and acceptors (S and Zn, respectively) are also given in the figure.

When gallium phosphide had been doped with copper [279], which acts as an acceptor in GaP, very high values of the resistivity were obtained at room temperature (up to 10^{10} $\Omega \cdot cm$) which, in the opinion of Allen and Cherry [279], were due to the strongly compensated state of the samples. The activation energy for electrical conduction in these samples was found to be 0.68 eV. The Hall mobility of holes in copper-doped samples was about 100 $cm^2 \cdot V^{-1} \cdot sec^{-1}$ at a density of $2 \cdot 10^{17}$ cm^{-3}.

The photoconductivity was discovered in these samples at wavelengths of 5800 Å and 1.87 μ.

According to the published data, the electron mobility does not exceed 110-130 $cm^2 \cdot V^{-1} \cdot sec^{-1}$ [279, 298, 299]. This value should be considerably larger since there is no reason for assuming that the electron mobility in GaP should be less than the hole mobility.

Gallium phosphide, like aluminum phosphide, has a wide forbidden band and interesting luminescent properties. GaP crystals emit, on passing dc or ac through them, orange radiation; the activator is probably gallium. An addition of zinc or copper makes the luminescence dark red. The luminescence is also observed if a layer of GaP powder is placed between electrodes made of metal and conducting glass [290].

Zinc-doped crystals exhibit electroluminescence in the regions of 6250 and 5650 Å. In investigating the dependence of the photoconductivity on the wavelength, excitation lines were found at ~4200 and 5600 Å (in undoped samples) and ~4200 Å and 6000 Å (in zinc-doped crystals) [288]. Rectification at a point contact (p- and n-type samples) has also been observed [288], the ratio of the forward and reverse currents sometimes reaching 10^5.

An investigation of the optical absorption of gallium phosphide has shown the presence of an absorption band in the wavelength range 1-4 μ; the width of the band increased with the electron density (this was not observed in p-type samples) [186]. Similar experiments have been used to determine the permittivity: ε = 8.5 ± 0.2 in the wavelength range 1-12 μ, and ε = 10.2 at wavelengths longer than 40 μ. The carrier multiplication at a p-n junction in gallium phosphide has also been investigated.

It is suggested by Gross et al. [286] that the valence band of GaP consists of several overlapping sub-bands.

Table 4 lists the principal properties of gallium phosphide.

Applications of Gallium Phosphide.

Reports have been published of the preparation of GaP diodes, which can work at temperatures above 500°C [271]. Diodes have been prepared capable of withstanding large reverse voltages of up to 135 V. The most detailed data on laboratory samples of GaP semiconducting devices [271] indicate the possibility of preparing diodes and transistors from this compound even in the present state of its technology. GaP diodes have been made by establishing an ohmic contact using liquid Ga—In alloy (t_{mp}~ 16°C) and a rectifying contact in the form of a sharp tungsten wire. The reverse currents in such diodes were ~ $3 \cdot 10^{-8}$ A under voltages of 5-10 V. Diodes such as these can rectify currents up to 0.5 A, at temperatures up to 800°C. The establishment of two point contacts in series has produced characteristics of the type met in point-contact transistors with high base resistance and a current amplification factor of $\alpha \approx 1$.

Numerous investigations of the luminescence of GaP indicate that this material may be suitable for the preparation of semiconductor light sources.

COMPOUNDS BASED ON INDIUM

Two indium-based compounds of the $A^{III}B^{V}$-type — indium antimonide and arsenide — are already widely used in semiconductor technology. They exhibit exceptionally high carrier mobilities, which govern their remarkable properties. InSb and InAs are among the few semiconducting materials which are used in the mass production of semiconducting devices [36, 24, 79].

In contrast, very little is known about the other two compounds of this group — indium phosphide, InP, and particularly, indium nitride, InN. Nevertheless, using the relationships governing the properties of similar semiconducting materials, one can predict a number of important properties for the phosphide and the nitride, which eventually may be used in the manufacture of semiconducting devices. Indium phosphide cannot be regarded as a refractory material since its melting point is 1060°C. However, at its melting point the phosphorus vapor pressure is high, and the technique for preparing indium phosphide is quite complicated; therefore, we thought it was desirable to consider the properties of this compound in the present book.

Indium Nitride InN

Indium nitride has not been investigated much, and singularly little is known about its semiconducting properties. The phase diagram of the indium—nitrogen system has not been investigated at all and very few properties of this compound are known at present.

TABLE 5. Principal Properties of Indium Compounds

Compound	T_{mp}, °C	Heat of fusion, kcal/mole	Heat of formation, kcal/mole	Structure	Lattice parameter, Å	Density, g/cm³	Microhardness, kg/mm²	Thermal expansion coefficient, deg⁻¹	Forbidden band width, eV, at 0°K	Electron mobility, cm²·V⁻¹·sec⁻¹ at 20°C	Hole mobility, cm²·V⁻¹·sec⁻¹ at 20°C	Refractive index	Absorption band edge, μ	Magnetic susceptibility	Effective electron mass
InN	1200 [227]	—	4.6 [29]	Wurtzite	a = 3.533, c = 5.693, c/a = 1.611 [29]	6.88 [29]	—	—	—	—	—	—	—	—	—
InP	1058 ± 3 At p= 21 ± 5 atm [293]	15.2 [301]	—	Sphalerite	5.869 [302, 300]	—	410–430 [303, 302]	$4.5 \cdot 10^{-6}$ [295]	1.34 [295, 304]	3400 (for $n=10^{17}$ cm⁻³) [295], 5000 (for $n = 6 \cdot 10^{15}$ cm⁻³) [305]	50 (for $n= 6 \cdot 10^{15}$ cm⁻³) [295]	3.0 [295], 3.3 [297]	1.0 [297]	$-45.6 \cdot 10^{-6}$	$(0.077 \pm 0.005)\, m_0$ [308]

Preparation of Indium Nitride

Indium nitride has been prepared only by indirect methods. The heating of the double salt of ammonium—indium fluoride $(NH_4)_3InF_6$ in a current of ammonia NH_3 at 580°C [230] and 630°C [270] produces a black soft powder InN. The same product has been obtained in the decomposition of indium chloride monoammoniate $InCl_3 \cdot NH_3$, brought about by boiling on a hot substrate [227]. The attempt to obtain InN directly from nitrogen and indium has not succeeded. This is because [73] InN is thermodynamically unstable at temperatures and pressures relatively close to normal. A change in temperature and pressure may make it possible for In and N to combine directly.

The Growing of Indium Nitride Single Crystals

The possibility of growing indium nitride single crystals has not yet been investigated. The vapor pressure at the melting point of indium nitride should be very high, at least not less than that for indium phosphide (20-25 atm). Therefore, such methods of growing of single crystals as the Bridgman or Czochralski methods (using stoichiometric melts) can hardly be used. The most promising is the gas-transport reaction method, and the method of growing single crystals from solutions (cf. the first part of the book).

Properties of Indium Nitride

Indium nitride has been obtained only in the form of soft, black fairly dense (~ 7 g/cm^3) [29], powder, having the wurtzite structure and a melting point of ~ 1200°C [227]. The heat of formation of indium nitride is, according to [29], 4.6 kcal/mole. It is stable at room temperature and begins to oxidize at 600°C; its dissociation becomes marked at 620°C in vacuum [227].

The available data on the physical properties of indium nitride are not very reliable. It has been reported [270] that InN exhibits metallic conduction (electrical resistance of the order of $4 \cdot 10^{-3}$ $\Omega \cdot$ cm, with a temperature coefficient of the resistance $+ 0.00037$ deg^{-1}).

The electrical resistance of InN has been measured using bars made from pressed powder [270]. It is very likely that the samples contained metallic indium, which led to these low values of the resistance.

Approximate theoretical calculations of the forbidden band width show that InN should be a typical semiconductor with a forbidden band $\Delta E = 2.4$ eV wide.

Table 5 lists some data on the properties of indium nitride.

Applications of Indium Nitride

If indium nitride has the wide forbidden band suggested for it, its pure single crystals should be transparent in the visible part of the spectrum, beginning from 5000 Å. Indium nitride may possibly be used for optical filters, to cut off the whole short-wavelength part of visible light and the near ultraviolet wavelengths as well. Indium nitride should also have electroluminescent properties, which may find technical application.

Indium Phosphide InP

Indium phosphide has been investigated more than any other compound, except gallium phosphide, dealt with in the present book. The great interest in this compound is due to its very valuable properties, listed in Table 5.

Preparation of Indium Phosphide

Indium phosphide was first synthesized by heating metallic indium in a current of phosphorus mixed with hydrogen [309]. Subsequently, other methods have been used. For example, indium phosphide may be obtained in the form of a powder by thermal decomposition (at 500°C) of a complex compound which precipitates on passing phosphine PH_3, carried in a stream of nitrogen, through indium monoiodide [273]. Indium phosphide powder may be obtained also by the exchange phosphorization reaction (800-900°C):

$$2In + Zn_3P_2 \rightarrow 2InP + 3Zn$$

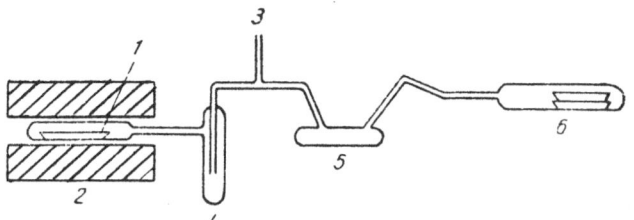

Fig. 57. Schematic representation of a system used to synthesize indium phosphide: 1) boat; 2) furnace; 3) tube; 4,5) reservoirs; 6) reaction chamber.

After this reaction, the excess of Zn_3P_2 and Zn is driven off in vacuum at 1000°C [243]. It has also been suggested that indium phosphide can be obtained in powder form at relatively low temperatures (up to 300°C) by the reduction of organic compounds of indium with phosphine [310]. The low temperature of such a process is a great technical advantage.

All these methods of preparing indium phosphide do not yield the large single crystals or polycrystalline aggregates that might be suitable for measuring the electrical properties of the compound. The powder obtained has to be melted, which may contaminate it, apart from posing the problem of one's having to deal with the relatively high vapor pressure of phosphorus at the congruent melting point.

Another disadvantage of the methods which yield powder are the large quantities of secondary products, which may introduce undesirable impurities into the final product.

In view of this, there is considerable interest in methods of direct fusion of indium and phosphorus in stoichiometric or nonstoichiometric ratio. The simplest method is to fuse these elements in a quartz ampoule placed in a single-temperature furnace [215, 302]. However, the ingots obtained in this way always contain several percent of excess indium which must be removed somehow, for example, by directional crystallization. Another interesting way of preparing this substance is a two-temperature method of synthesis in a horizontal ampoule, using a charge in which there is a deficiency of phosphorus compared with the stoichiometric ratio [311, 84, 312]. Because of this, the vapor pressure of the phosphorus above this type of melt is much lower than that above a stoichiometric melt, and this makes it easier to carry out the process.

The subsequent use of directional cooling makes it possible to drive off the excess indium and to crystallize the melt in a polycrystalline (or, sometimes, a single-crystal) ingot. Such directional cooling may be produced either by the motion of the ampoule or by the motion of the crystallization front, keeping the ampoule fixed. This is done by the slow cooling of the furnace, which produces a temperature gradient.

There is also great interest in the method of synthesizing indium phosphide simultaneously with the purification of the initial components in the same apparatus in which the synthesis is carried out [313]. Figure 57 illustrates schematically the system used to purify phosphorus and indium before the reaction chamber is placed in a furnace.

The preparation stage before synthesis is carried out as follows: an indium charge is placed in the upper of the two boats in a chamber 6; the charge is melted on evacuating the system through 3; then, molten indium drains away to the lower boat through an aperture in the upper boat, and an oxide layer, always present on indium, remains in the upper boat. In this way, oxygen, which enters the melt with the indium, is removed. Then, red phosphorus, placed in boat 1, is sublimated by a heater 2 at 300-380°C and is driven into a reservoir 4; next, the whole system is sealed (by sealing tube 3). The phosphorus, now present in the reservoir 4, is converted from red to white by the rapid cooling of the vapor and is then driven into the reservoir 5 at 150°C, and hence into the reaction chamber 6. Before distillation takes place from the reservoir 5 to the chamber 6, the reservoir 4 is sealed off and this is followed by the sealing off of the reservoir 5. Thus, the initial materials needed in the synthesis are finally brought together in the reaction chamber; they are not exposed to air after purification. The ratio of the constituents is taken to ensure an excess of indium. Thus, InP crystallizes from a solution in In [313].

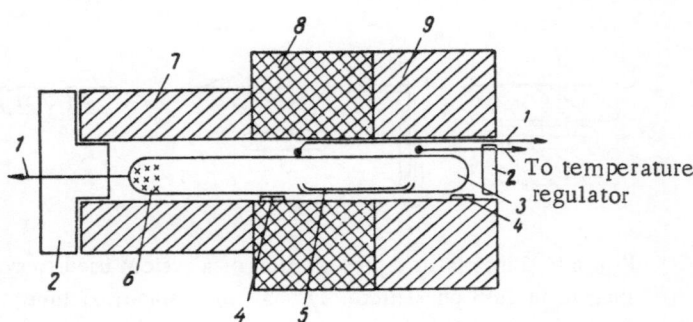

Fig. 58. Furnace used to synthesize indium phosphide: 1)
thermocouples; 2) clinker plug; 3) quartz tube; 4) alundum
support for tube; 5) boat with indium; 6) phosphorus; 7)
heater for phosphorus; 8) low-temperature heater for indium;
9) high-temperature heater for indium.

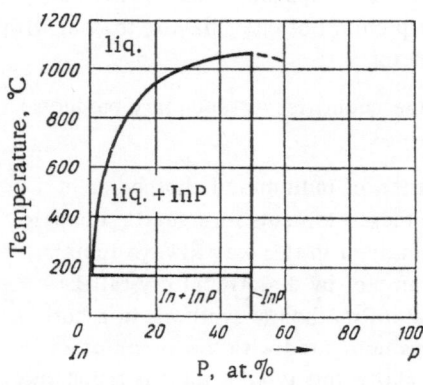

Fig. 59. Phase diagram of the indium—
phosphorus system.

Fig. 60. Order of indium and phosphorus
atomic layers in indium phosphide in
the direction [111].

After these operations, the synthesis may be carried out,
for example, in a horizontal triple-heater apparatus, such as
illustrated in Fig. 58.

A constant temperature gradient 1000-1060°C (the higher
temperature is slightly lower than the melting point of indium
phosphide) is established by two of the heaters (8 and 9) having
separate windings. After the stabilization of the gradient along
the ampoule, the heater 7 raises the temperature of the phosphorus, present at the cold end of the ampoule, to 485°C.

The phosphorus vapor saturates the In until the concentration of the phosphorus in the indium reaches a level which corresponds to 1000°C liquidus temperature, i.e., to the minimum
temperature in the ampoule. In accordance with the phase
diagram, further dissolving of the phosphorus in the melt precipitates solid InP at the cold end of the ampoule and sets in
motion the crystallization front in the direction of the hot end
of the ampoule. The excess of indium, always present in the
melt, is driven to the hot end of the ampoule, where it finally
freezes. Such directional crystallization removes a number of
impurities from the InP and simultaneously produces large crystallites in a polycrystalline ingot, and sometimes even single
crystals (15 × 5 × 2 mm).

The preparations of a molten solution of phosphorus in
indium and its slow crystallization is a much simpler method
of synthesizing indium phosphide [101]. This method makes it
possible to heat safely an ampoule — particularly if it is also
shaken during the heating — and the process may be carried out at relatively low temperatures, exceeding slightly
higher than the liquidus temperature for alloys with 10-20 at.% P. This method gives directly single-crystal
samples of indium phosphide up to 12 mm long and 1.5 mm thick. The crystals are pulled out mechanically
from an excess of indium and then they are etched in a solution of hydrochloric acid in glycerine heated to 180°C. *

_____

*Glycerine raises the boiling point of the etchant and dilutes the acid.

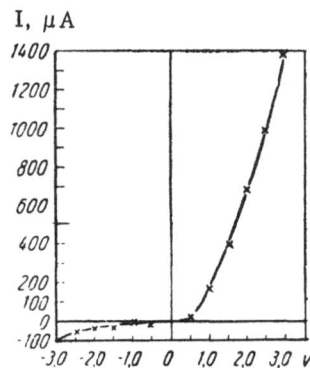

Fig. 61. Current—voltage characteristic of p-n junctions in InP.

After traces of excess indium are removed, pure single-crystal samples are obtained on which electrical measurements can be carried out.

It is worth mentioning that an excess of indium (or phosphorus) in the preparation of indium phosphide does not impair the electrical properties of the final product. This is due to the negligible solubility of In and P in solid indium phosphide [302] and the low electrical activity of In and P impurities in indium phosphide. This phenomenon is characteristic of all $A^{III}B^{V}$-type substances.

Of the listed methods of preparing indium phosphide, those involving the directional crystallization of a concentrated solution of InP in In are of greatest interest because of their relative simplicity and safety. The synthesis and crystallization of molten solutions of InP in In is extremely simple and may be carried out in any laboratory. Moreover, such a synthesis gives directly small, but very perfect, indium phosphide single crystals on which some electrical investigations may be carried out. The same method can be used to dope indium phosphide with any admixture, directly during the molten stage. The distribution of impurities introduced in this way should be more uniform than in any other doping method.

The Growing of Indium Phosphide Single Crystals

Some methods of preparing indium phosphide single crystals have already been described in connection with the synthesis of this material. As previously stated, sometimes the synthesis is simultaneous with the growth of single crystals, which is very convenient from the point of view of semiconductor technology. We shall consider next the methods of preparing single crystals in which the initial material is indium phosphide synthesized by another method.

The method of pulling single crystals from an InP—In molten solution (the Czochralski method) is very promising [313]. The use of a nonstoichiometric melt (60 at.% In and 40 at.% P) makes it possible to carry out this process at relatively low phosphorus vapor pressures (~5 atm) in the working chamber.

Indium phosphide single crystals may be prepared also by the gas-transport reaction method. For indium phosphide, such a process may be described by the reversible reaction equation

$$InP + InI_3 \rightleftharpoons InI + P.$$

The reaction has the tendency to shift to the left as the temperature falls and conversely. Therefore, InP single crystals grow at the cold end of the ampoule [105] in the form of correctly faced plates, attaining a size of 1 mm or more.

The most promising method of preparing large indium phosphide single crystals is, in the present author's opinion, the Czochralski method of pulling from a molten InP—In solution, since by suitable modification, this method should yield large oriented single crystals. However, to obtain a uniform distribution of impurities along a single crystal, additional treatment will be needed, as in the growth of single crystals of other materials (for example, germanium) by this method.

Properties of Indium Phosphide

Physicochemical Properties. It can be regarded as established that InP is the only compound of the In—P system.

The phase diagram of the In—P system has been investigated several times but only in the region from In to InP (from 0 to 50 at.% P).

Figure 59 shows the phase diagram of the In—P system [83].

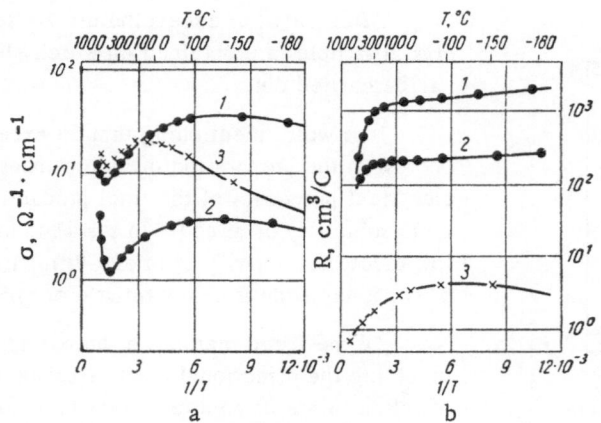

Fig. 62. Temperature dependence of the electrical conductivity (a) and of the Hall coefficient (b) in indium phosphide samples: 1, 2) n-type samples; 3) p-type samples.

Other investigations [314, 301] have shown that the nature of the phase diagram given in [83] is correct but the vapor pressure at the melting point (60 atm) given in [83] is obviously too high. The quartz apparatus used in [83] would have hardly stood phosphorus vapor pressure of the order of 40 atm. Moreover, somewhat later [215, 302] it was reported that indium phosphide, very close to the stoichiometric composition, had been prepared in a standard quartz ampoule in a single-temperature furnace.

The most reliable data on the congruent melting temperature of indium phosphide and its dissociation vapor pressure at this temperature are given in [293]. These quantities are 1058 ± 3°C and 21 ± 5 atm, respectively.

Indium phosphide is quite stable in air and begins to oxidize only at 500°C. Its dissociation in vacuum becomes noticeable at 750°C. Indium phosphide is not dissolved by dilute acids and alkalis; it is etched only by strong reagents (mixtures of concentrated HCl and HNO_3, i.e., aqua regia) [84].

Indium phosphide crystallizes in the sphalerite lattice with a lattice period of 5.86 Å [302, 300].

X-ray diffraction investigations of platelet crystals, obtained from dilute solutions of P in In, have shown that the most developed faces are the (111) planes and, because the compound consists of two types of atom and these atoms alternate in the lattice in pairs (as shown in Fig. 60), the platelets obtained have indium atoms on one side (side A) and phosphorus atoms on the other (side B). Thus, indium phosphide, like other substances of the $A^{III}B^{V}$-type, is a polar crystal [315, 316]. This should be borne in mind in the preparation of indium phosphide single crystals. On the phosphorus side, the crystallization is easier than on the indium side, as in the growth of InSb single crystals [317]. This is due to the higher activity of the B side, which appears, for example, in the etching of single-crystal plates.

Indium phosphide is usually etched in aggresive media due to the high stability of this substance; for example, the visualization of grain boundaries in a polycrystalline sample requires pure aqua regia or a 50% solution of aqua regia in the water.

Dislocations are revealed by etching in HNO_3 : HCl (1:1) [316] or in a 0.4 N solution of Fe^{3+} in concentrated HCl [315]. Dislocation etch pits, in the form of triangles or hexagons, usually appear only on the phosphorus side, because the indium side simply becomes very rough.

Practically nothing is known about the segregation coefficients of impurities in InP. The known values of the effective segregation coefficients are given below.

Impurity	Ge	Sn	S	Se
Segregation coefficient	0.05	0.03	0.8	0.6
Reference	[318]	[318]	[84]	[84]

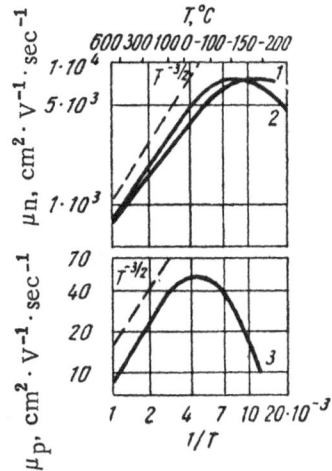

Fig. 63. Temperature dependence of the carrier mobility in indium phosphide.

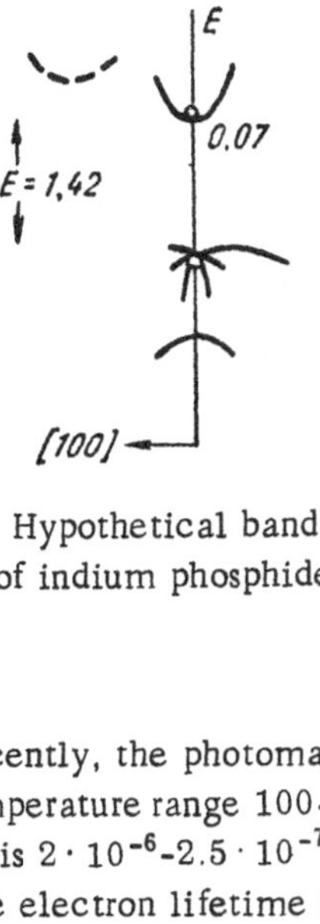

Fig. 64. Hypothetical band structure of indium phosphide.

The segregation coefficients of Cd and Zn acceptors are very small [84].

Physical Properties. In investigating the process of self-diffusion in InP, it has been found that In diffuses faster than P. The temperature dependence of the diffusion coefficients obeys the exponential law $D = D_0 \exp(-E/kT)$, where, for In,

$$D_0 = 1 \cdot 10^5 \text{ cm}^2/\text{sec}; E = 3.85 \pm 0.05 \text{ eV};$$

and for P

$$D_0 = 7 \cdot 10^{10} \text{ cm}^2/\text{sec}; E = 5.65 \text{ eV}.$$

Weiser [319] used the different solubility of impurities in InP and In to form p-n junctions in single crystals by a very original method: a sample was subjected to local fusion; at the heated spot, the chemical compound dissociated and phosphorus vaporized. Consequently, the impurities were redistributed and a p-n junction was formed at the heated spot, while the remaining indium crystallized. This automatically produced a good electrical contact. The local heating was carried out in a helium atmosphere using a 900°C helium jet.

Figure 61 shows the current—voltage characteristic of a p-n junction in InP prepared in this way. Polycrystalline samples, prepared by pulling from the melt [313], had an electron density of 10^{17} cm^{-3} at room temperature and relatively high mobility $\mu_H = 3000$ cm$^2 \cdot$ V$^{-1} \cdot$ sec^{-1}. Figure 62 shows the temperature dependences of the electrical conductivity and of the Hall coefficient. From the $R(T)$ and $\sigma(T)$ dependences, the temperature dependences of the electron and hole mobilities were derived (Fig. 63).

An even higher electron mobility (5000 cm$^2 \cdot$ V$^{-1} \cdot$ sec^{-1}) was obtained for indium phosphide single crystals having a carrier density of $6 \cdot 10^{15}$ cm^{-3} [305]. Indium phosphide exhibits greater transparency than germanium in the infrared part of the spectrum [313]; this is a convenient feature.

Recently, the photomagnetic effect and photoconductivity of n-type indium phosphide were investigated in the temperature range 100-300°K [320]. This investigation showed that the minority carrier lifetime at t = 300°K is $2 \cdot 10^{-6}$-$2.5 \cdot 10^{-7}$ sec for holes and $1.7 \cdot 10^{-3}$-$2.2 \cdot 10^{-3}$ sec for electrons. On cooling the InP to 100°C the electron lifetime rises by a factor of 3-6, while the hole lifetime becomes shorter.

The details of the energy band structure of indium phosphide are not fully known at present. Measurements of the piezoresistance of n-type InP, having an electron density $2.5 \cdot 10^{16}$ cm^{-3}, at 77 and 300°K have shown that the conduction band of this substance is spherical (the minimum of the conduction band lies at the center of the Brillouin zone) [330].

The recently published [331] investigations of the optical absorption at 6, 20, and 77°K of n-type InP samples 7-170 μ thick (no substrate), having an electron density of $5 \cdot 10^{15}$ cm^{-3}, show that the forbidden band width is 1.420 eV at 6°K and 1.413 eV at 77°K. The temperature coefficient of the forbidden band width above 100°K is $\beta = -2.9 \cdot 10^{-4}$ eV/deg, which is in serious disagreement with the previously published data [332].

Figure 64 shows a hypothetical band structure of InP [79]. Some data on the properties of indium phosphide are given in Table 5.

Applications of Indium Phosphide

The semiconducting properties of indium phosphide are similar to those of such well-known semiconductors as silicon and gallium arsenide. This similarity allows us to predict, with certainty, that InP may be successfully employed to prepare a variety of semiconducting devices: diodes, photocells, transistors. There are reports of the observation of stimulated emission from InP [333]. Unfortunately, there is as yet no information on the ionization energies of impurities in InP but, by analogy with other $A^{III}B^{V}$ compounds, we may assume that group II acceptors and group VI donors will have ionization energies not greater than 0.1 eV. Hence, we may conclude that InP devices can work over a wide range of temperatures: from -100 to $+200$ or $+250°C$. The high electron mobility leads us to expect good frequency characteristics of semiconducting devices made of InP. Indium phosphide may also be used as an infrared radiation receiver in the impurity photoconductivity region.

LITERATURE CITED

1. N. A. Goryunova, Gray Tin, Author's Abstract of Dissertation for the Degree of Candidate of Technical Sciences, Leningrad, IONKh (1950).
2. H. Welker, Z. Naturforsch. 7a:744 (1952).
3. A. F. Ioffe, Physics of Semiconductors, Leningrad, Izd. Akad. Nauk SSSR (1957).
4. M. Faraday, Experimental Researches in Electricity, Vol. I [Russian translation], Moscow, Izd. Akad. Nauk SSSR (1947), pp. 413-419.
5. K. Wagner, Z. Phys. Chem. 21:42 (1933).
6. K. Wagner and W. Schottky, Z. Phys. Chem. (B) 11:467 (1930).
7. J. Bardeen and W. Shockley, Phys. Rev. 80:72 (1950).
8. W. Shockley, Electrons and Holes in Semiconductors [Russian translation], Moscow, IL (1953).
9. E. M. Conwell, Proc. Inst. Radio Engrs. 46:1281 (1958).
10. S. H. Koenig, R. D. Brown, and W. Schillinger, Phys. Rev. 128:1668 (1962).
11. E. M. Conwell and V. E. Weisskopf, Phys. Rev. 77:388 (1950).
12. C. Erginsoy, Phys. Rev. 79:1013 (1950).
13. D. L. Dexter and F. Seitz, Phys. Rev. 86:964 (1952).
14. J. R. Haynes, Phys. Rev. 98:1866 (1955).
15. J. R. Haynes and H. B. Briggs, Phys. Rev. 86:647 (1952).
16. J. R. Haynes and W. C. Westphal, Phys. Rev. 101:1676 (1956).
17. R. Braunstein, Phys. Rev. 99:1892 (1955).
18. T. S. Moss and T. D. H. Hawkins, J. Phys. Radium 17:712 (1956).
19. R. N. Hall, Phys. Rev. 87:387 (1952).
20. W. Shockley and W. T. Read, Phys. Rev. 87:835 (1952).
21. É. I. Adirovich, Some Problems in the Theory of Luminescence of Crystals, Moscow, Gostekhteoretizdat (1950).
22. V. Roberts and J. E. Quarrington, J. Electron. 1:152 (1955).
23. T. S. Moss, Optical Properties of Semiconductors, New York, Academic Press Inc. (1959) [Russian translation, Moscow, IL (1961)].
24. R. A. Smith, Semiconductors, New York, Cambridge University Press (1959) [Russian translation, Moscow, IL (1962)].
25. W. C. Dash and R. Newman, Phys. Rev. 99:1115 (1955).
26. S. Zwerdling, B. Lax, and L. Roth, Phys. Rev. 108:1402 (1957); 109:2207 (1958).
27. H. B. Briggs and R. C. Fletcher, Phys. Rev. 87:1130 (1952).
28. W. Keiser, R. J. Collins, and H. Y. Fan, Phys. Rev. 91:230, 1380 (1952).
29. R. Juza and H. Hahn, Z. Anorg. Allgem. Chem. 239:282 (1938); 244:133 (1940).
30. Ya. G. Dorfman, Dokl. Akad. Nauk SSSR 81:765 (1951).
31. W. Shockley, Phys. Rev. 90:491 (1953).
32. R. N. Dexter, H. J. Zeiger, and B. Lax, Phys. Rev. 104:637 (1956).
33. E. Burstein, G. S. Picus, and H. A. Gebbie, Phys. Rev. 103:825 (1956).
34. Yu. I. Ukhanov, Fiz. Tverd. Tela 4:2739 (1962); 5:108 (1963); Yu. I. Ukhanov and Yu. V. Mal'tsev, Fiz. Tverd. Tela 4:3215 (1962).
35. G. Busch and U. Winkler, Determination of the Characteristic Parameters of Semiconductors [Russian translation], Moscow, IL (1959).
36. N. B. Hannay (ed.), Semiconductors, New York, Reinhold Publishing Corp. (1959). [Russian translation, Moscow, IL (1962)].

37.   Ya. A. Fedotov and Yu. V. Shmartsev, Transistors, Moscow, Izd. "Sov. Radio" (1960).

38.   Yu. P. Maslakovets and V. K. Subashiev (eds.), Semiconducting Radiant-Energy Converters, Collection of Translations into Russian, Moscow, IL (1959).

39.   V. I. Stafeev, New Principles of Operation of Semiconducting Devices, Author's Abstract of Dissertation for Doctorate of Physicomathematical Sciences, Moscow (1961).

40.   A. Shavlov, S. Fogel', and L. Dalbredzher, Optical Quantum Generators (Lasers), Moscow, IL (1962).

41.   W. Shockley, Proc. Inst. Radio Engrs. 46:954 (1958).

42.   R. K. Mueller, J. Appl. Phys. 30:1004 (1959).

43.   H. A. Gebbie, P. C. Banbury, and C. A. Hogarth, Proc. Phys. Soc. London 63B:371 (1950).

44.   Ya. A. Fedotov, Crystal Triodes (Transistors), Moscow, Gosénergoizdat (1955).

45.   G. C. Decay and I. M. Ross, Bell System Tech. J. 34:1149(1955).

46.   D. A. Jenny, Proc. Inst. Radio Engrs. 46:717, 959 (1958).

47.   W. Hartel, Siemens-Z. 28:376 (1954).

48.   V. P. Zhuze, "Technical Applications of the Hall Effect," in collection: Semiconductors in Science and Technology, Vol. I, Leningrad, Izd. Akad. Nauk SSSR (1957).

49.   I. T. Sheftel', "Semiconducting Thermistors," in collection: Semiconductors in Science and Technology, Vol. I, Leningrad, Izd. Akad. Nauk SSSR (1957).

50.   V. V. Aleksandrov, V. I. Pruzhinina, A. I. Rekov, T. S. Tarakanova, and E. A. Teplov, Fiz. Tverd. Tela 1:1587 (1959).

51.   J. A. Becker, Trans. AIME 65:711 (1946).

52.   S. F. Jacobs, Electronics, April 1, 72 (1960).

53.   R. A. Smith, Semiconductors [Russian translation], Moscow, IL (1962). See reference 24.

54.   R. H. Bube, Photoconductivity of Solids, New York, John Wiley & Sons, Inc. (1960) [Russian translation, Moscow, IL (1962)].

55.   Ya. A. Fedotov (ed.), Semiconducting Devices and Their Applications, Moscow, Izd. "Sov. Radio" (1956-1963), Nos. 1-7.

56.   W. Beyen, P. Bratt, H. Davis, L. Johnson, H. Levinstein, and A. Mackay, J. Opt. Soc. Am. 49:686 (1959).

57.   Q. A. Morton, M. L. Schultz, and W. E. Harty, RCA Rev. 20:599 (1959).

58.   D. M. Chapin, C. S. Fuller, and G. L. Pearson, Bell Lab. Record 33:241 (1955).

59.   H. P. R. Frederikse and R. F. Blunt, Proc. Inst. Radio Engrs. 43:1828 (1955).

60.   D. A. Jenny, J. J. Lofersky, and P. Rappaport, Phys. Rev. 101:1208 (1956).

61.   N. G. Basov, O. N. Krokhin, and Yu. M. Popov, Collection: p-n Junctions in Semiconductors, Tashkent, Izd. Akad. Nauk UzSSR (1962), p. 93.

62.   M. G. A. Bernard and G. Durrafourg, Phys. Status Solidi 1:699 (1961).

63.   M. I. Nathan, W. P. Dumke, G. Burns, F. H. Dill, and G. Lasher, Appl. Phys. Letters 1:62 (1962).

64.   T. M. Quist, R. H. Rediker, R. J. Keyes, W. E. Krag, B. Lax, A. L. McWhorter, and H. J. Zeigler, Appl. Phys. Letters 1:91 (1962).

65.   E. M. Biron, Zh. Russ. Fiz.-Khim. Obshchestva 47:964 (1915).

66.   S. A. Shchukarev, Zh. Obshch. Khim. 24:582 (1954).

67.   N. A. Goryunova, Chemistry of Diamond-Like Semiconductors, Leningrad, Izd. Leningrad. Gos. Universiteta (1963).

68.   N. A. Goryunova, Investigations in Chemistry of Semiconductors, Author's Abstract of Dissertation for Doctorate of Chemical Sciences, Moscow, IONKh (1958).

69.   J. P. Suchet, J. Phys. Chem. Solids 16:265 (1960); 21:156 (1961).

70.   N. Selar, J. Appl. Phys. 33:2999 (1962).

71.   I. S. Asanabe, Mem. Fac. Sci., Kyushu Univ. Ser. B 2:82 (1956).

72.   B. Stone and D. Hill, Phys. Rev. Letters 4:282 (1960).

73.   B. F. Ormont, Collection: Problems in Metallurgy and Physics of Semiconductors, Izd. Akad. Nauk SSSR (1961), p. 5.

74.   P. Manga, J. Phys. Chem. Solids 20:268 (1961).

75.   D. P. Belotskii, Collection: Problems in Metallurgy and Physics of Semiconductors, Izd. Akad. Nauk SSSR (1961).

76.   L. Kleinman and J. C. Phillips, Phys. Rev. 117:460 (1960).

77. R. Taylor and C. A. Coulson, Proc. Phys. Soc., London A65:834 (1952).

78. J. Lagrenaudie, J. Chem. Phys. 53:222 (1956).

79. C. Hilsum and A. C. Rose-Innes, Semiconducting III-V Compounds, New York, Pergamon Press, Inc. (1961) [Russian translation, Moscow, IL (1963)].

80. R. Kauer and A. Rabenau, Z. Naturforsch. 12a:942 (1957).

81. P. Popper, Progr. Dielectrics 1:219 (1959).

82. Z. A. Willey, J. Metals 8:263 (1956).

83. J. van den Boomgard and K. Schol, Philips Res. Rept. 12:127 (1957).

84. O. G. Folberth, Halbleiterprobleme, Braunschweig, No. 5 (1960).

85. F. A. Kröger and D. Nobel, J. Electron. 1:190 (1955).

86. W. G. Spitzer, M. Gershenzon, C. J. Frosch, and D. F. Gibbs, J. Phys. Chem. Solids 11:339 (1959).

87. A. N. Krestovnikov and V. N. Vigdorovich, Chemical Thermodynamics, Moscow, Metallurgizdat(1961).

88. H. E. Buckley, Crystal Growth New York, John Wiley & Sons Inc. (1951) [Russian translation, Moscow, IL (1954)].

89. V. D. Kuznetsov, Crystals and Crystallization, Moscow, Gostekhteoretizdat (1954).

90. S. Honigman, Growth and Form of Crystals [Russian translation], Moscow, IL (1961).

91. A. R. Verma, Crystal Growth and Dislocations [Russian translation], Moscow, IL (1958).

92. P. W. Bridgman, Proc. Am. Acad. Arts Sci. 60:305 (1925).

93. J. T. Edmund, R. F. Broom, and F. A. Cunnell, Services Electronics Res. Lab. Tech. J. 6:123 (1957).

94. L. Roth and W. E. Taylor, Proc. Inst. Radio Engrs. 40:1338 (1952).

95. D. A. Petrov (ed.), Collection: Germanium [Russian translation], Moscow, IL (1955).

96. E. Billig and D. Gasson, J. Sci. Instr. 35:360 (1958).

97. W. G. Pfann, Zone Melting,New York, John Wiley & Sons, Inc. (1958) [Russian translation, Moscow, Metallurgizdat (1960)].

98. R. Gremmelmaier, Z. Naturforsch. 11a:511 (1956).

99. J. L. Richard, J. Sci. Instr. 34:289 (1957).

100. J. M. Whelan and G. H. Wheatley, Bull. Am. Phys. Soc. 2:120 (1957).

101. G. Wolff, P. H. Keck, and J. D. Broder, Bull. Am. Phys. Soc. 29:16 (1954).

102. W. G. Pfann, Trans. AIME 203:961 (1955).

103. H. Schäfer, Chemische Transportreaktionen, Weinheim/Bergstr., Verlag Chemie (1961).

104. R Nitsche, H. U. Bölsterli, and M. Lichtensteiger, J. Phys. Chem. Solids 21:199 (1961).

105. G. R. Antell and D. Effer, J. Electrochem. Soc. 106:509 (1959).

106. G. R. Antell, Brit. J. Appl. Phys. 12:687 (1961).

107. V. S. Trofimov, Piroda, No. 5:25-29 (1941).

108. A. E. Fersman, Essays on the History of Rocks, Vol. I, Moscow, Izd. Akad. Nauk SSSR (1954).

109. I. I. Shafranovskii, Diamonds, Leningrad, Izd. Akad. Nauk SSSR (1953).

110. G. F. V. Smith, Gemstones, London, Methuen & Co. (1950).

111. S. Tolansky, Contemp. Phys. 1:96 (1959).

112. A. E. Fersman, Diamond, Moscow, Izd. Akad. Nauk SSSR (1957).

113. P. Freeman and H. A. van der Velden, Physica 18:1, 9 (1952).

114. Editorial note, Mining Journal 247:590 (1956).

115. F. P. Bundy, J. Chem. Phys. 88:631 (1963).

116. H. T. Hall, Rev. Sci. Instr. 31:125 (1960).

117. H. P. Bovenkerk, et al., Nature, London, 184:1094 (1959).

118. R. H. Wentorf and H. P. Bovenkerk, J. Chem. Phys. 36:1987 (1962).

119. G. A. Wolff, L. Toman, N. Y. Field, and J. C. Clark, Halbleiter und Phosphore, Braunschweig (1958), p. 463.

120. D. L. Bark and S. A. Friedberg, Phys. Rev. 111:1275 (1958).

121. M. Omar, N. S. Pandya, and S. Tolansky, Proc. Roy. Soc., London, A225:33 (1954).

122. N. S. Pandya and S. Tolansky, Proc. Roy. Soc., London, A225:40 (1954).

123. A. R. Patel and S. Tolansky, Proc. Roy. Soc., London, A243:41 (1957).

124. S. Tolansky, The Microstructure of Diamond Surfaces, London, N. A. G. Press (1955).

125. R. Robertson, J. F. Fox, and A. E. Martin, Phil. Trans. Roy. Soc., London, Ser. A 232:463 (1934).

126.   K. G. Ramanathan, Proc. Indian Acad. Sci., Sect. A 24:137 (1946).

127.   D. E. Blackwell and G. B. B. Sutherlend, J. Chem. Phys. 46:9 (1949).

128.   W. Keiser and W. L. Bond, Phys. Rev. 115:857 (1959).

129.   J. F. H. Custers, Physica 18:489 (1952); 20:183 (1954).

130.   H. B. Dyer and P. T. Wedepohl, Proc. Phys. Soc., London, B69:410 (1956).

131.   I. G. Austin and R. Wolfe, Proc. Phys. Soc., London, B69:329 (1956).

132.   C. Allemand and J. Rossel, Helv. Phys. Acta 27:519 (1954).

133.   C. C. Klick and R. J. Mourer, Phys. Rev. 76:179 (1949).

134.   K. G. McKay, Phys. Rev. 74:1606 (1950).

135.   A. G. Redfield, Phys. Rev. 94:526 (1954).

136.   P. T. Wedepohl, Proc. Phys. Soc., London, B70:177 (1957).

137.   R. T. Bate and R. K. Willardson, Proc. Phys. Soc., London, 74:363 (1959).

138.   P. J. Kemmey and E. W. J. Mitchell, Proc. Roy. Soc., London, A263:420 (1961).

139.   M. D. Bell and W. J. Leivo, Phys. Rev. 111:1227 (1958).

140.   A. Halperin and J. Nahum, J. Phys. Chem. Solids 18:297 (1961).

141.   E. W. J. Mitchell, J. Phys. Chem. Solids 8:444 (1959).

142.   C. D. Clark et al., Phil. Mag. 5:127 (1960).

143.   C. E. Ryan, in: Silicon Carbide (J. R. O'Connor and J. Smiltens [eds.]), London, Pergamon Press (1960), p. 15.

144.   L. A. Patrick, J. Appl. Phys. 28:765 (1957).

145.   H. Baumhauer, Z. Krist. 55:249 (1915).

146.   N. W. Thibault, Am. Mineralogist 29:249 (1944).

147.   J. A. Lely, Ber. Deut. Keram. Ges. 32:229 (1955).

148.   J. A. Lely and F. A. Kroger, Proceedings of the Garmisch Conference (1956).

149.   D. R. Hamilton, J. Electrochem. Soc. 105:735 (1958).

150.   H. C. Chang and L. J. Kroko, Conf. Paper 57-1131, AIEE, Chicago (1957).

151.   D. R. Hamilton, in: Silicon Carbide (J. R. O'Connor and J. Smiltens [eds.]), London, Pergamon Press (1960), p. 43.

152.   J. Drowart and G. de Maria, in: Silicon Carbide (J. R. O'Connor and J. Smiltens [eds.]), London, Pergamon Press (1960), p. 16.

153.   R. C. Ellis, in: Silicon Carbide (J. R. O'Connor and J. Smiltens [eds.]), London, Pergamon Press (1960), p. 124.

154.   R. N. Hall, J. Appl. Phys. 29:914 (1958).

155.   R. J. Scace and G. A. Slack, J. Chem. Phys. 30:1551 (1959).

156.   F. A. Halden, in: Silicon Carbide (J. R. O'Connor and J. Smiltens [eds.]), London, Pergamon Press (1960), p. 115.

157.   R. E. Honig, J. Chem. Phys. 22:1610 (1954).

158.   C. F. Powell, Campbell, and Gonser, Vapor-Plating, John Wiley & Sons, Inc., New York (1955).

159.   D. B. Lyon, C. M. Olson, and E. D. Lewis, Trans. Electrochem. Soc. 96:359 (1949).

160.   H. C. Thenrer, Bell Lab. Record 33:327 (1955).

161.   F. B. Litton and H. C. Andersen, J. Electrochem. Soc. 101:287 (1954).

162.   B. Rubin, G. H. Moates, and J. R. Weiner, J. Electrochem. Soc. 104:656 (1957).

163.   S. P. Kleshchevnikova, Ya. E. Pokrovskii, and W. I. Rumyantseva, Zh. Tekhn. Fiz. 27:1645 (1957).

164.   P. H. Keck, in: Silicon Carbide. (J. R. O'Connor and J. Smiltens [eds.]), London, Pergamon Press (1960), p. 130.

165.   W. Brenner, in: Silicon Carbide (J. R. O'Connor and J. Smiltens [eds.]), London, Pergamon Press (1960), p. 110.

166.   H. Nowotny et al., Monatsh. Chem. 85:255 (1954).

167.   W. V. Wright and F. Bartels, in: Silicon Carbide (J. R. O'Connor and J. Smiltens [eds.]), London, Pergamon Press (1960), p. 31.

168.   Humphrey et al., US Bur. Mines Rept. Invest. 4888 (1952).

169.   Harman and Muxer, USAEC, Report BMI 748 (1952).

170.   A. I. Miklashevskii, Chemical Properties and Analysis of Carborundum, Moscow, ONTI (1939).

171. P. Schwarzkopf and R. Kieffer, Refractory Hard Metals, The Macmillan Company, New York (1953).

172. A. Taylor and R. Jones, in: Silicon Carbide (J. R. O'Connor and J. Smiltens [eds.]), London, Pergamon Press (1960), pp. 147-154.

173. G. Bosch, Philips Res. Rept. 16:455 (1961).

174. J. L. Birman, in: Silicon Carbide (J. R. O'Connor and J. Smiltens [eds.]), London, Pergamon Press (1960), pp. 257-280.

175. H. R. Philipp, Phys. Rev. 111:440 (1958).

176. H. R. Philipp and E. A. Taft, in: Silicon Carbide (J. R. O'Connor and J. Smiltens [eds.]), London, Pergamon Press (1960), pp. 366-370.

177. S. Kobayasi, J. Phys. Soc. Japan 13:261 (1958).

178. W. J. Choyke and L. Patrick, Phys. Rev. 105:1721 (1957); in: Silicon Carbide (J. R. O'Connor and J. Smiltens [eds.]), London, Pergamon Press (1960), pp. 306-311.

179. M. Namba, J. Phys. Chem. Solids 2:339 (1957).

180. M. Schön, Z. Naturforsch. 8a:442 (1953).

181. J. H. Racette, Phys. Rev. 107:1542 (1957).

182. R. W. Keyer, in: Silicon Carbide (J. R. O'Connor and J. Smiltens [eds.]), London, Pergamon Press (1960), pp. 395-398.

183. H. J. van Daal, C. A. A. J. Greeble, W. F. Knippenberg, and H. J. Vink, J. Appl. Phys., Suppl. 32:2225 (1961).

184. M. Mirzabaev, V. M. Tuchkevich, and Yu. V. Shmartsev, Izv. Akad. Nauk SSSR, Ser. Fiz. 28:1300 (1964).

185. L. Patrick and W. J. Choyke, in: Silicon Carbide (J. R. O'Connor and J. Smiltens [eds.]), London, Pergamon Press (1960), pp. 281-305.

186. L. Patrick and W. J. Choyke, J. Appl. Phys. 30:236 (1959).

187. R. G. Pohl, in: Silicon Carbide (J. R. O'Connor and J. Smiltens [eds.]), London, Pergamon Press (1960), pp. 312-330.

188. G. Busch and H. Labhart, Helv. Phys. Acta 19:463 (1946).

189. Hung-Chi Chang, Z. Le May, and L. F. Wallace, in: Silicon Carbide (J. R. O'Connor and J. Smiltens [eds.]), London, Pergamon Press (1960), pp. 496-507.

190. O. V. Losev, Telegrafiya i Telefoniya, No. 18:61 (1923); No. 26:403 (1924); No. 44:485 (1927); No. 53:153 (1929).

191. O. W. Lossew [O. V. Losev], Wireless World 15:93 (1924); Z. Fernmeld. Techn. No. 7:97 (1926).

192. W. T. Eriksen, in: Silicon Carbide (J. R. O'Connor and J. Smiltens [eds.]), London, Pergamon Press (1960), pp. 376-383.

193. K. Thissen and G. Jungk, Phys. Status Solidi 2:473 (1962).

194. D. Hofman, J. A. Lely, and J. Volger, Physica 23:236 (1957).

195. W. G. Spitzer et al., Phys. Rev. 113:127, 133 (1959), in: Silicon Carbide (J. R. O'Connor and J. Smiltens [eds.]), London, Pergamon Press (1960), pp. 347-365.

196. V. V. Pasynkov, in collection: Semiconductors in Science and Technology, Vol. I, Leningrad, Izd. Akad. Nauk SSSR (1957), pp. 314-337.

197. N. P. Bogoroditskii, V. V. Pasynkov, G. F. Kholuyanov, and D. A. Yas'kov, Izv. Akad. Nauk SSSR, Ser. Fiz. 20:1571 (1956).

198. T. C. Taylor, in: Silicon Carbide (J. R. O'Connor and J. Smiltens [eds.]), London, Pergamon Press (1960), pp. 431-442.

199. T. E. Kharlamova and G. F. Kholuyanov, Fiz. Tverd. Tela 2:426 (1960).

200. M. I. Iglitsyn, M. Mirzabaev, V. M. Tuchkevich, E. F. Fedotova, and Yu. V. Shmartsev, Fiz. Tverd. Tela 6:2673 (1964).

201. J. W. Faust, in: Silicon Carbide (J. R. O'Connor and J. Smiltens [eds.]), London, Pergamon Press (1960), pp. 403-419.

202. C. Goldberg and J. W. Ostrowski, in: Silicon Carbide (J. R. O'Connor and J. Smiltens [eds.]), London, Pergamon Press (1960), pp. 453-461.

203. Jäger and Westenbrink, Proc. Acad. Sci. Amsterdam 29:1218 (1926).

204. E. Friedrich and L. Sittig, Z. Anorg. Chem. 143:293 (1925).

205.   R. H. Wentorf, J. Chem. Phys. 26:956 (1957).

206.   R. H. Wentorf, J. Chem. Phys. 34:809 (1961); 36:1990 (1962).

207.   G. V. Samsonov, Boron, Its Compounds and Alloys, Kiev, Izd. Akad. Nauk UkrSSR (1960).

208.   R. C. Vickery, Nature, London, 184:268 (1959).

209.   E. Wiberg and H. Michand, Z. Naturforsch. 96:497 (1954).

210.   W. Kroll, Z. Anorg. Chem. 102:17 (1918).

211.   P. Popper and T. A. Ingles, Nature, London, 179:1075 (1957).

212.   S. Rundquist, XVI Congrès International de Chimie Pure et Appliquée, Paris, 1957; Mémoire présente a la section de chimie minéral., p. 539.

213.   J. Perry, S. Laplace, and B. Post, Acta Cryst. 11:310 (1958).

214.   F. V. Williams and R. A. Ruehrwein, J. Am. Chem. Soc. 82:1330 (1960).

215.   A. S. Borshchevskii and D. N. Tret'yakov, Fiz. Tverd. Tela 1:1483 (1959).

216.   M. A. Besson, Compt. Rend. 113:78 (1891).

217.   H. Moissan, Compt. Rend. 113:624, 726 (1891).

218.   Yu. A. Valov and É. Yu. Lubenskaya, Proceedings of the Twentieth Scientific Conference of the Leningrad Structural Engineering Institute, Physics Section, Leningrad (1962), p. 31.

219.   S. Rundquist, Acta Chem. Scand. 16:1 (1962).

220.   B. D. Stone, Chem. Abstr. 56:4194 (1962).

221.   E. P. Stambaugh, J. F. Miller, and R. C. Himes, Metallurgy of Elemental and Compound Semiconductors, New York, Interscience Publishers (1961), p. 317.

222.   B. D. Stone and R. A. Ruehrwein, Chem. Abstr. 56:2971 (1962).

223.   V. J. Matkovich, Acta Cryst. 14:93 (1961).

224.   B. F. Ormont, Zh. Neorgan. Khim. 4:2176 (1959).

225.   Yu. A. Klyachko, Zh. Prikl. Khim. 14:84 (1941).

226.   N. Dudzinski, J. Inst. Metals 83:444 (1955).

227.   T. Renner, Z. Anorg. Allgem. Chem. 298:22 (1959).

228.   A. Addamiano, J. Electrochem. Soc. 108:1072 (1961).

229.   M. Tetsuo and T. Yasaku, J. Phys. Soc. Japan 15:203 (1960).

230.   A. Busev, Uch. Zap. Leningr. Gos. Ped. Inst. 29:303 (1940).

231.   H. Ott, Z. Physik. 22:201 (1924).

232.   M. V. Stackelberg and K. Speess, Z. Phys. Chem. A175:127 (1935).

233.   F. Wolff, Z. Anorg. Chem. 83:109, 125, 161 (1913).

234.   A. E. Vol, Structure and Properties of Binary Metal Systems, Vol. I, Moscow, Fizmatgiz (1959).

235.   O. Kubaschewski and E. L. Evans, Metallurgical Thermochemistry [Russian translation], Moscow, IL (1954), p. 246.

236.   R. Fischer and V. Oesterheld, Z. Electrochemie 21:50 (1915).

237.   R. Röntgen and H. Braun, Metallwirtschaft 11:459 (1932).

238.   French Patents, Nos. 367124, 415252 (1910); German Patent, No. 241339 (1910).

239.   N. G. Anslie, S. E. Blum, and J. E. Woords, J. Appl. Phys. 33:2391 (1962).

240.   Selected Values of Chemical Thermodynamic Properties, Natl. Bur. Std. (U.S.), Circ. 500, Washington, D. C. (1952).

241.   B. Grashchenko, V. Darovskii, and A. Zhand Zh'yan, Tr. Vses. Alyumin. Magnievyi Inst., No. 16:90 (1937).

242.   P. Schwarzkopf and R. Kieffer, Refractory Hard Metals [Russian translation], Moscow, Metallurgizdat (1957).

243.   A. Addamiano, J. Am. Chem. Soc. 82:1537 (1960).

244.   G. V. Samsonov and L. L. Vereikina, Phosphides, Kiev, Izd. Akad. Nauk UkrSSR (1961).

245.   M. G. Grimmeis and W. Kischio, J. Phys. Chem. Solids 16:302 (1960).

246.   V. A. Presnov, M. A. Krivov, V. N. Vertoprakhov, A. G. Grigor'eva, and E. V. Malisova, in collection: Problems in Metallurgy and Physics of Semiconductors, Moscow−Leningrad, Izd. Akad. Nauk SSSR (1959), p. 127.

247.   Yu. I. Pashintsev, Author's Abstract of Dissertation for Candidate of Physicomathematical Sciences, Minsk State University (1959).

248. G. Natta and L. Passerini, Gazz. Chim. Ital. 58:458 (1928).

249. N. A. Goryunova, Zh. Vses. Khim. Obshchestva im. D. I. Mendeleeva 5:522 (1960).

250. F. Herman, J. Electron. 1:103 (1955).

251. M. E. Levina, Vestn. Mosk. Gos. Univ., Ser. Mat. Mekhan. Astron. Fiz. i Khim.,p. 241 (1956).

252. F. Wöhler, Ann. Chem. 87:146 (1827).

253. V. Goldschmidt, Skrifter Norske Videnskaps-Akad. Oslo, II: Mat.-Naturv. Kl., No. 8 (1927).

254. Fonzec-Diacon. Compt. Rend. 130:1314 (1900).

255. T. Numata, J. Phys. Soc. Japan 17:878 (1962).

256. I. I. Burdiyan, Solid Solutions of the AlSb—GaSb System and Their Electrical Properties, Author's Abstract of Dissertation for Candidate of Physicomathematical Sciences, Leningrad, Physicotechnical Institute (1959).

257. W. Köster and B. Thoma, Z. Metallk. 46:291 (1955).

258. C. J. Frosch and L. Derick, J. Electrochem. Soc. 108:251 (1961).

259. A. S. Borshchevskii and D. N. Tret'yakov, Proceedings of the Twentieth Conference of the Leningrad Structural Engineering Institute, Physics Section, Leningrad (1962), p. 14.

260. V. N. Romanenko and G. V. Nikitina, Izv. Akad. Nauk SSSR, Otd. Tekhn. Nauk, Met. i Topliva, No. 3:56 (1962).

261. J. P. Suchet, Chimie Physique des Semiconducteurs, Paris, Dunod (1962).

262. H. G. Grimmeis, R. Groth, and J. Maak, Z. Naturforsch. 15a:799 (1960).

263. H. G. Grimmeis and H. Koelmans, Z. Naturforsch. 14a:264 (1959).

264. W. C. Johnson, J. B. Parsons, and M. Grew, J. Phys. Chem. Solids 36:2651 (1952).

265. G. A. Wolff, I. Adams, and J. W. Mellichamp, Phys. Rev. 114:1262 (1959).

266. G. S. Zhdanov and G. M. Mirman, Zh. Éksperim. i Teor. Fiz. 6:1201 (1936).

267. R. Juza, Naturforschung und Medizin in Deutschland 1939-1949, p. 24; Z. Anorg. Chem. 2:61 (1948).

268. M. R. Lorenz and B. B. Binkovsky, J. Electrochem. Soc. 109:24 (1962).

269. Klemm, Jacobi, and Tilk, Z. Anorg. Chem. 207:187 (1932).

270. R. Juza and A. Rabenau, Z. Anorg. Chem. 285:212 (1956).

271. D. Mandelkorn, Proc. Inst. Radio Engrs. 47:2012 (1959).

272. V. Goldschmidt, Strukturberichte, p. 136 (1931).

273. D. Effer and G. R. Antell, J. Electrochem. Soc. 107:110 (1960).

274. G. V. Samsonov, L. L. Vereikina, and Yu. B. Titkov, Zh. Prikl. Khim. 35:242 (1962).

275. M. Gershenzon and R. M. Mikulyak, J. Electrochem. Soc. 108:548 (1961).

276. R. H. Keck and J. D. Broder, Phys. Rev. 90:521 (1953).

277. G. A. Kalyuzhnaya, D. N. Tret'yakov, A. S. Borshchevskii, and A. A. Vaipolin, in collection: Research in Semiconductors—New Semiconducting Materials, Kishinev, Kartya Moldovenyaské (1964), p. 123. [English translation: Soviet Research in New Semiconductor Materials, New York, Consultants Bureau (1965), p. 80.]

278. C. J. Frosch, M. Gershenzon, and D. F. Gibbs, Symposium on the Growth of III—V Compounds, Batelle Memorial Institute, Columbus, Ohio, 1959.

279. J. W. Allen and R. J. Cherry, J. Phys. Chem. Solids 23:163, 509 (1962).

280. G. F. Alfrey and C. S. Wiggins, Z. Naturforsch. 15a:267 (1960).

281. M. Gershenzon and R. M. Mikulyak, Solid State Electronics (1963).

282. M. Gershenzon and R. M. Mikulyak, J. Appl. Phys. 32:1338 (1961).

283. D. A. Kleinman and W. G. Spitzer, Phys. Rev. 118:110 (1960).

284. H. W. Allison, J. Appl. Phys. 34:231 (1963).

285. F. G. Ulman, J. Electrochem. Soc. 109:805 (1962).

286. E. F. Gross, G. K. Kalyuzhnaya, and D. S. Nedzvetskii, Fiz. Tverd. Tela 4:3543 (1962).

287. E. F. Gross and D. S. Nedzvetskii, Dokl. Akad. Nauk, No. 5:146 (1962).

288. H. G. Grimmeis and H. Koelmans, Philips Res. Rept. 15:290 (1960).

289. S. Iizima and M. Kikuchi, J. Phys. Soc. Japan 16:1783 (1961).

290. G. A. Wolff, R. O. Hebert, and J. D. Broder, Phys. Rev. 100:1144 (1955).

291. D. A. Petrov (ed.), New Methods of Preparation of Semiconductor Single Crystals [Russian translation], Moscow, IL (1962).

292.   T. Renner, Solid-State Electron. 1:41 (1960).

293.   D. Richman, J. Phys. Chem. Solids 24:1131 (1963).

294.   M. Rubinstein, J. Electrochem. Soc. 109:69C (1962).

295.   H. Welker and H. Weiss, in collection: New Semiconducting Materials [Russian translation], Moscow, IL (1958), p. 35.

296.   O. G. Folberth, Z. Naturforsch. 9a:1050 (1954).

297.   F. Oswald and E. Schade, Z. Naturforsch. 9a:611 (1954).

298.   D. N. Nasledov and S. V. Slobodchikov, Fiz. Tverd. Tela 4:2775 (1962).

299.   C. J. Frosch, M. Gershenzon, D. F. Gibbs, and W. G. Stitzen, J. Phys. Chem. Solids 11:339 (1959).

300.   G. Giesecke and H. Pfister, Acta Cryst. 11(5):369 (1958).

301.   M. Shafer and K. Weiser, J. Phys. Chem. 61:1424 (1957).

302.   N. A. Goryunova, N. N. Fedorova, and V. I. Sokolova, Zh. Tekhn. Fiz. 28:1672 (1958).

303.   A. S. Borshchevskii, N. A. Goryunova, and N. K. Takhtareva, Zh. Éksperim. i Teor. Fiz. 27:1048 (1957).

304.   K. Weiser, J. Phys. Chem. 61:511 (1957).

305.   M. Gliksman and K. Weiser, J. Electrochem. Soc. 105:728 (1958).

306.   G. Busch and R. Kern, Helv. Phys. Acta 32:24 (1959).

307.   M. Matyas, Czech. J. Phys. 9:257 (1959).

308.   E. D. Palik and R. F. Wallis, Phys. Rev. 123:131 (1961).

309.   A. Tiel and W. Koelisch, Z. Anorg. Chem. 66:288 (1910).

310.   R. Didchenko, US Patent 3,010,792/cl. 23-204; (US Patent Office, 1961), XI, Vol. 772, No. 4, 1099.

311.   O. G. Folberth and H. Weiss, Z. Naturforsch. 10a:615 (1955).

312.   R. Guire and K. Weiser, US Patent 2,871,100/cl. 23-204; (US Patent Office, 1955), No. 523, 718.

313.   T. C. Harman, J. I. Genco, W. P. Allred, and H. H. Goering, J. Electrochem. Soc. 105:731 (1958); see also reference [56].

314.   W. Köster and W. Ulrich, Z. Metallk. 49:365 (1958).

315.   H. C. Gatos and M. C. Lavine, J. Electrochem. Soc. 107:427 (1960).

316.   H. Pfister, Z. Naturforsch. 16a:427 (1961).

317.   H. C. Gatos, P. Z. Moody, and M. C. Lavine, J. Appl. Phys. 31:212 (1960); see also reference [56].

318.   O. G. Folberth and E. Schillmann, Z. Naturforsch. 12a:943 (1957).

319.   K. Weiser, J. Appl. Phys. 29:299 (1958).

320.   M. P. Mikhailova, D. N. Nasledov, and S. V. Slobodchikov, Fiz. Tverd. Tela 4:1227 (1962).

321.   Z. S. Medvedeva and G. D. Mitkina, Preparation of Boron Phosphide and Arsenide and Some of Their Properties, Proceedings of the All-Union Conference on Semiconducting Compounds, held December 18-23, 1961 (Abstracts of Papers), Moscow–Leningrad, Izd. Akad. Nauk SSSR (1961).

322.   Electronics, 33(35):78 (1960).

323.   W. J. Choyke, D. R. Hamilton, and L. Patrick, Phys. Rev. 133:A1163 (1964).

324.   F. Bassani and M. Yoshimine, Phys. Rev. 130:20 (1963).

325.   A. R. Verma, Crystal Growth and Dislocations, London, Butterworths (1953); cf. also reference [91].

326.   H. J. McSkimin and W. L. Bond, Phys. Rev. 105:116 (1957).

327.   R. S. Krishnan, Proc. Indian Acad. Sci., Sect. A 24:33 (1946).

328.   H. R. Philipp and E. A. Taft, Phys. Rev. 136:A1445 (1964).

329.   W. B. Wilson, Phys. Rev. 127:1549 (1962).

330.   A. Sagar, Phys. Rev. 117:101 (1960).

331.   W. J. Turner et al., Phys. Rev. 136:A1467 (1964).

332.   F. Oswald, Z. Naturforsch. 10a:927 (1955); 9a:181 (1954).

333.   K. Weiser and R. S. Levitt, Appl. Phys. Letters 2:178 (1963).

334.   R. J. Archer, R. Y. Koyama, E. E. Loebner, and R. C. Lucas, Phys. Rev. Letters 12:538 (1964).

335.   C. C. Wang, M. Cardona, and A. G. Fischer, RCA Rev. 25:159 (1964).

336.   A. S. Borshchevskii, G. A. Kalyuzhnaya, A. D. Smirnova, N. K. Takhtareva, and D. N. Tret'yakov, Izv. Akad. Nauk SSSR, Ser. Fiz. 28:985 (1964).

337.   J. W. Faust and H. F. John, J. Phys. Chem. Solids 25:1407 (1964).

338.   J. D. Broder and G. A. Wolff, J. Electrochem. Soc. 110:1150 (1963).

339.   W. W. Piper and S. J. Polich, J. Appl. Phys. 32:1278 (1961).
340.   H. Flicker, B. Goldstein, and P. A. Hoss, J. Appl. Phys. 35:2959 (1964).
341.   M. I. Iglitsyn et. al., Fiz. Tverd. Tela 6:2673 (1964).
342.   R. Zallen and W. Paul, Phys. Rev. 134:A1628 (1964).

INDEX